Anonymus

# Catalogue of Fruit Trees, Roses, Shrubs, Forest Trees, Bulbs and Plants

Anonymus

**Catalogue of Fruit Trees, Roses, Shrubs, Forest Trees, Bulbs and Plants**

ISBN/EAN: 9783742812247

Manufactured in Europe, USA, Canada, Australia, Japa

Cover: Foto ©Klaus-Uwe Gerhardt /pixelio.de

Manufactured and distributed by brebook publishing software
(www.brebook.com)

Anonymus

# Catalogue of Fruit Trees, Roses, Shrubs, Forest Trees, Bulbs and Plants

# TO OUR CUSTOMERS.

*Though we passed through another period of drought during the early summer, thanks to copious rains in July and careful cultivation, our Nursery Stock has made good growth and is of splendid quality.*

*We have made further reductions in many of our prices, and are confident that everything offered is of excellent value.*

*We supply nothing but British Grown Fruits and Roses, which have in the past proved themselves of far better value than those imported. We also claim that Trees and Plants grown in our Eastern Counties Nurseries possess a hardier constitution, and will hold their own anywhere when transplanted.*

*APPLES, PEARS and PLUMS are more and more being grown both in private gardens and commercially. Our stock is exceptionally good.*

*CURRANTS, RASPBERRIES and other small fruit; no garden should be without a few bushes, which give a quick return, and are invaluable for dessert, cooking and jam-making.*

*ROSES have always been a great feature with us, and this year are better than ever. Every individual Rose is staked and allowed to develop from the start.*

*FOREST TREES. We have over a million of these for sale.*

*ORNAMENTAL TREES and SHRUBS. Owing to the drought, our stock of these is short, and we shall not be able to replace. Please order early.*

## BULBS and FORCING ROOTS.

*Our Collection of HYACINTHS, TULIPS and DAFFODILS includes all the newest varieties, as well as well-known older ones. We are pleased to be able to again reduce many of our prices to meet as far as possible the competitive inferior stuff which is offered from abroad.*

# APPLES.

We are this season offering many varieties of fruit trees and bushes at a further reduction on last year's prices. When it is remembered that we only send out trees of the very highest quality, we are confident that intending planters will decide at once to take advantage of these reduced prices and complete their schemes of new fruit plantations and the filling of vacant places in the fruit garden.

The Apple is now generally recognized as the staple fruit for all general purposes, and the greatly increased demand for these trees bide fair to be maintained for some years. In spite of the great demand of last season, and the especially dry Summer just experienced, we are able to offer a very fine range of Apple Trees in Standards, Half and Three-quarter Standards, Bushes, Pyramids, Cordons, and Horizontal Trained Trees. To enable our customers to select their requirements we have set out the Apples in sections, showing the Varieties of each style that we can offer with the prices attached.

The best time for planting Fruit Trees is Autumn, as early as convenient, in October and November, after the young trees have shed their leaves, and while the soil retains some of the summer warmth. Planting may however be carried out any time to the end of March if the weather is open.

We work our Apples on both the Crab and broad-leaved Paradise Stock, according to the varieties. Dwarf Apples on the Paradise Stock are of especial value for garden planting, and have come into favour during recent years. They are much dwarfer in growth, come into bearing and profit much sooner, are easier to thin and spray, and they produce almost continuously abundant crops of much finer fruit. For small holdings, or where the tenure is uncertain, they are specially recommended.

To small holders, and those who have but recently become interested in fruit growing, we offer the advice of our experts as to the best positions, soils, and varieties, and we invite correspondence. All enquiries will be promptly dealt with.

## PRUNING FRUIT TREES.

The future success of Fruit Trees, assuming that they are carefully planted, depends largely upon the pruning of the trees. From our very wide correspondence and from personal observation we are convinced that the importance of this work is not generally appreciated. Many failures and much disappointment would be avoided if the work was properly undertaken and at the right time.

We can with confidence recommend a little book entitled, "The Handy Book on Pruning," written by Mr. Jas. Udale (Chief Horticultural Instructor for Worcestershire), and we shall be pleased to supply copies post free for 3/-.

## CORDON APPLES.

A very useful type of tree where space is limited as they can be planted 18 inches apart on wall or arch.

We have a very fine lot of Single Cordons, for list of varieties see pages 5 to 7, marked with an *.

APPLE.
*Single Cordon Trained
Growing at our Nurseries.*

# APPLES.
## FULL STANDARDS.

### STANDARD APPLES ON STOUT STEMS.

We offer below a choice selection of Standard Apples including most of the best Dessert and Kitchen varieties. These trees are most suitable for Orchards where the trees are required to give large quantities of fruit, they can be planted 24 to 30 feet apart. In forming young trees it is wise to shorten the shoots two-thirds their length the first season, and one-third the second season, then a tree with a good foundation is secured.

Ard Cairn Russet (D). Dec. to March.

Barnack Beauty (D). Dec. to March.

Beauty of Bath (D). August.

Blenheim Orange (D.K.). Dec. to Feb.

Bramley's Seedling (K). Dec. to April.

Christmas Pearmain (D). Dec. and Jan.

Cox's Orange Pippin (D). Nov. to March.

Crimson Bramley (K). Dec. to April.

Dr. Harvey (K). Oct. to March.

Dumelow's Seedling (K). Nov. to March.

Ecklinville Seedling (K). Sept. and Oct.

Ellison's Orange (D). Oct.

Feltham Beauty (D). Sept.

Grenadier (K). Oct. and Nov.

Herring's Pippin (D). Sept. and Oct.

Lady Sudeley (D). Aug. and Sept.

Lord Derby (K). Nov. and Dec.

Mere de Menage (K). Oct. to Jan.

Monarch (New) (K). Oct. to April. 10/6 each.

Mr. Gladstone (D). July and August.

Newton Wonder (K). Nov. to May.

Norfolk Beauty (K). Oct. to Dec.

Peasgood's Nonsuch (D.K.). Sept. to Jan.

Rev. W. Wilks (K). Sept. to Nov.

Stirling Castle (K). Oct. and Nov.

Vicar of Beighton (K). Dec. to May.

Warner's King (K). Oct. to Dec.

Worcester Pearmain (D.K.). Sept. and Oct.

### Each 7/6; per dozen 84/-

D. *denotes Dessert.* K. *Kitchen.*    *For descriptions of these Varieties see pages 5 to 7.*

# APPLES.

## SPECIAL OFFER.

### HALF AND THREE-QUARTER STANDARDS.

We give below a list of varieties of Apples of which we are able to supply half and three-quarter standards, and we especially recommend these to those who desire to grow trees on stems.

Where the land is to be cropped between the trees they will be found most suitable; they are also admirable for planting in ground which is used for poultry runs, the branches being a sufficient distance from the ground to prevent the birds destroying the fruit.

The trees are all well grown, of fruiting size, and will quickly give a return. The varieties include all the best for cooking and dessert, both early and late sorts.

These trees should be planted 18 ft. apart.

Annie Elizabeth. Dec. to April.
Ard Cairn Russet (D). Dec. to March.
Beauty of Bath (D). August.
Blenheim Orange (D.K.). Dec. to Feb.
Bramley's Seedling (K). Dec. to April.
Charles Ross (D). November.
Christmas Pearmain (D). Dec. and Jan.
Cox's Orange Pippin (D). Nov. to Mar.
Crimson Bramley (K). Dec. to April.
Dr. Harvey (K). Oct. to March.
Ellison's Orange (D). October.
Ecklinville Seedling (K). Sept. & Oct.
Early Victoria (K). Aug. and Sept.
Feltham Beauty (D). September.
Gascoigne's Scarlet (K). Nov. to Jan.
Herring's Pippin (D). Sept. and Oct.
Keswick Codlin (K). Aug. and Sept.
King of Pippins (D). Nov. and Dec.
Lady Sudeley (D). Aug. and Sept.
Lord Suffield (K). Aug. and Sept.
Lord Derby (K). Nov. and Dec.
Mere de Menage (K). Oct. to Jan.
Monarch, New (K). Oct. to April.
Newton Wonder (K). Nov. to May.
Norfolk Beauty (K). Oct. to Dec.
Red Victoria (K). Aug. and Sept.
Stirling Castle (K). Oct. and Nov.
The Queen (K). Oct. to Dec.
Warner's King (K). Oct. to Dec.

|  | Each | Doz. |
|---|---|---|
| Three-quarter Standards | 6/6 | 72/- |
| Half Standards | 5/6 | 60/- |

### HORIZONTAL TRAINED TREES.

For producing good crops of splendid quality there is no better style of tree than those especially trained for Espaliers. These trees may be planted along the sides of the path round the Kitchen Garden or Shrubbery walks, and they will form delightful features in Spring when in bloom and also furnish fruit of first-class quality. The trees should be planted 12—15 ft. apart, about 4 ft. from the path, and at once trained to canes or wires. The most suitable varieties are those mentioned below, which have been selected for training after careful trial.

Those marked with an * can be supplied in 2 Tiers, and those marked with a † can be supplied in 3 Tiers.

*Allington Pippin. Nov. to Jan.
*Annie Elizabeth (K). Dec. to April.
†*Beauty of Bath (D). July and Aug.
*Bismarok (K). Oct. to Dec.
†*Christmas Pearmain (D). Dec. and Jan.
*Cox's Orange Pippin (D). Nov. to March.
†*Crimson Bramley (K). Dec. to April.
†Ecklinville Seedling (K). Sept. and Oct.
†*Herring's Pippin (D). Sept. and Oct.
*King of Pippins (D). Oct. to Dec.
†*Lane's Prince Albert (K). Dec. to April.
†Lady Henniker (K). Nov. to Jan.
*Lady Sudeley (D). Aug. and Sept.
†*Lord Derby (K). Nov. and Dec.
†*Lord Suffield (K). Aug. and Sept.
†*Norvic. November.
*Peasgood's Nonsuch (D.K.). Sept. to Jan.
†*Rev. W. Wilks (K). Sept. to Nov.
†St. Everard (D). September.
*Vicar of Beighton (K). Dec. to May.
†Warner's King (K). Oct. to Dec.
†Worcester Pearmain (D.K.). Sept. and Oct.

*For descriptions of these Varieties see following pages.*

## 2 Tiers 7/6 each
## 3 Tiers 10/6 each

We also can supply a few extra sized Trees, 4 and 5 Tiers. Varieties and prices on application

*If desired we can supply permanent " Acme " Labels for all Fruit Trees. For prices see page 64.*

# APPLES—General List of Select Varieties.

### D. denotes Dessert.    K. Kitchen.

| | | |
|---|---|---|
| BUSHES .. .. .. | 4/- each | 42/- doz. |
| ,, Extra Strong Selected | 5/- ,, | 54/- ,, |
| PYRAMIDS .. .. .. | 7/6 ,, | 84/- ,, |
| CORDONS, Single .. .. | 5/- ,, | 54/- ,, |
| ,, Strong Fruiting .. | 6/- ,, | 66/- ,, |

In all cases where customers leave the selection of varieties to us, they may rely on only good trees of the best kinds being supplied. It is most important, however, that the style or shape of trees required should be clearly stated when ordering.

Those marked with an * can be supplied as Single Cordons.

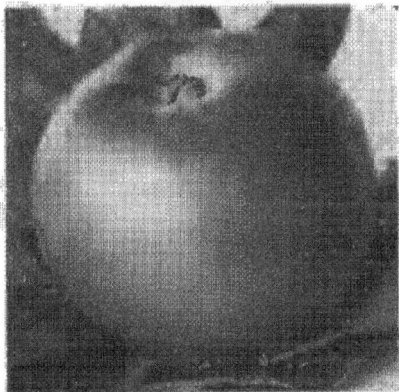

BRAMLEY'S SEEDLING.

**ALLINGTON PIPPIN.** A splendid medium-sized Apple introduced a few years ago, and which has taken a position in the front rank as a first-rate dessert variety and a reliable bearer. In form it resembles Cox's Orange. Nov. to Jan.

**ANNIE ELIZABETH** (K). Very fine late Apple, excellent keeping qualities. Dec. to April.

**ARD CAIRN RUSSET** (D). Medium size fruit, round, conical. Colour dark red, nearly covered with pale russet. Flesh yellow, fine good russet flavour. Dec. to March.

**AROMATIC RUSSET** (D). A first-class Russet of good flavour, and a great favourite in some localities. Nov. to Jan.

**BARNACK BEAUTY** (D). Medium, oval, regular. Colour golden yellow, dark red flush and faint stripes. Flesh crisp, fair flavour, briskly acid, yellowish. Dec. to March.

* **BEAUTY OF BATH** (D). A very handsome early variety, has a brisk, sub-acid flavour. July and Aug.

**BISMARCK** (K). A very fine large Apple from New Zealand ; one of the best varieties in cultivation for market or the private garden. Of great size, brilliant colour, a most profuse bearer. Oct. to Dec.

* **BLENHEIM ORANGE** (D.K.). Well-known and splendid variety ; large handsome fruit. Dec. to Feb.

* **BRAMLEY'S SEEDLING** (K). A large handsome fruit, resembling Blenheim Orange. Dec. to April.

**CHRISTMAS PEARMAIN** (D). Fruit medium, oval ; colour dull yellow with brownish-red flush, slight russet and faint stripes. Flesh firm, yellowish, pleasantly flavoured. Dec. and Jan. Standards, Half and Three-quarter Standards only.

**CHARLES ROSS.** A superb dessert variety. A cross between " Cox's Orange" and " Peagood's Nonsuch." It has all the richness of flavour and handsome appearance of the parents ; an excellent bearer. Nov.

* **COX'S ORANGE PIPPIN** (D). Medium, of delicious flavour, finest dessert apple ; good habit, bears and grows well as Standard ; A1 as a garden tree, succeeding in all forms, but prefers a warm rich soil. Best flavoured from low trees on Paradise. Nov. to March.

* **COX'S POMONA** (D.K.). Fruit large, yellowish green, streaked all over with crimson, of pleasant flavour. Oct. and Nov.

* **CRIMSON BRAMLEY** (K). Similar in every respect except colour to Bramley's Seedling ; hardy, robust, heavy bearer. Dec. to April.

**DOCTOR HARVEY** (K). Very fine, large ; first-class for culinary uses. Oct. to March.

**DUMELOW'S SEEDLING** (K). Culinary Apple which keeps till March. Colour creamy yellow with scarlet flesh. Flesh crisp, white, acid. Standards only.

**EARLY VICTORIA** (K). Pale lemon-coloured, very early variety of the Codlin type ; very free bearer. Aug. and Sept.

**ECKLINVILLE SEEDLING** (K). A large and useful sort ; flesh white and tender ; a great bearer. Sept. and Oct.

* **ELLISON'S ORANGE** (D). A delicious dessert Apple approaching in appearance and flavour to Cox's Orange Pippin ; a good keeper. Oct.

**FELTHAM BEAUTY** (D). Fruit medium. Colour yellowish-green with red stripes and flush. Flesh yellowish, crisp, and very highly flavoured. Sept.

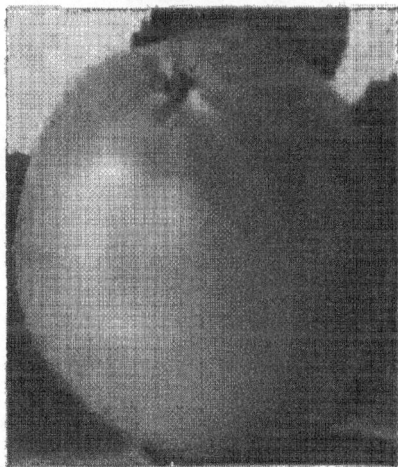

LORD SUFFIELD.

*If desired we can supply permanent " Acme " Labels for all Fruit Trees. For prices see page 64.*

## Apples (*continued*).

REV. W. WILKS.

**GASCOIGNE'S SCARLET** (K). Large, a distinct, richly coloured orchard apple, extremely handsome, and a great bearer; prolific on Paradise, and a healthy grower. Prune lightly. Nov. to Jan. F.C.R.H.S.

**GRENADIER** (K). Very large and prolific, flesh firm and juicy. Oct. and Nov.

**HARLING HERO** (K). New (Our Introduction). A most valuable addition to the late cooking varieties. It is a heavy cropper, a sturdy grower of a very hardy constitution, and thrives well on all soils. It possesses a remarkable beauty of colour and perfect shape; very large. Dec. to Feb. 3 years, 7/6 each.

**HERRING'S PIPPIN** (D). Deliciousand of handsome appearance, Pearmain shape, flesh yellow, crisp, and sweet. Standards, Cordons, and Pyramids. Sept. and Oct.

**IRISH PEACH** (D). Medium size, juicy and free; one of the best early Apples. August.

**JAMES GRIEVE** (D). Medium size; flavour of "Orange Pippin." Oct. to Dec.

**KING OF PIPPINS** (D). Fruit medium-sized; a richly flavoured and excellent dessert variety. Aug. and Sept.

**LANE'S PRINCE ALBERT** (K). Large, extremely handsome striped fruit, very prolific, keeps six months; its fertility is remarkable. Cordons or Bushes, on the Paradise Stock, produce grand exhibition fruit. F.C.R.H.S. Dec. to April.

**LADY HENNIKER** (K). Large fruit, golden yellow with bright red flush and a few broad broken stripes. Flesh tender, yellow. Blenheim flavour. Horizontal Trained Trees only.

**LADY SUDELEY** (D). Large fruit, bright golden yellow heavily splashed with brilliant scarlet stripes and flush. Flesh yellow, tender, very juicy, of delicious flavour. Aug. and Sept. Standards only.

**\*LORD DERBY** (K). Large, handsome, heavy cropper; one of the best. Nov. and Dec.

**LORD SUFFIELD** (K). Fine variety of "Keswick Codlin" type. Early and prolific; one of the best cooking Apples. Aug. and Sept.

**MÈRE DE MÉNAGE** (K). A handsome and useful culinary Apple; fruits are very large, of first-rate quality. Oct. to Jan. Standards only.

**MONARCH** (New). A cross between "Wellington" and "Peasgood's Nonsuch." Very large, firm fleshed, handsome appearance, quality superior, and requires less sugar. Heavy bearer and good sturdy grower. Oct. to April. 3 years, 7/6 each.

**MR. GLADSTONE** (D). Scarlet cheek, yellow flesh, a free bearer though not vigorous grower. Standards,and Horizontal Trained only. July and Aug.

**\*NEWTON WONDER** (K). Large; a handsome fruit, keeping soundly, free grower and bearer; one of the best. In growth and sturdiness this surpasses all others. A sterling kind for orchard or garden. F.C.R.H.S. Nov. to May.

**NORFOLK BEAUTY** (K). A cross between "Warner's King" and "Dr. Harvey." Fruit large, pale green changing to yellow. In appearance, intermediate between the two parents. Oct. to Dec.

**\*NORVIC, New** (Our Introduction). Large and handsome fruit, when ripe it is a greenish yellow; a first-rate cooking variety and of fine flavour; a most valuable acquisition both for private and market growers. Nov. 7/6 each.

**PEACEMAKER.** A very large and handsome culinary Apple. The fruit is pale yellow, flushed with crimson; a heavy cropper.

**\*PEASGOOD'S NONSUCH** (D.K.). Very large and handsome, pale yellow, with bright crimson. Of diffuse growth on Paradise; requires its roots well pruned to induce fertility in a young state. We recommend it for garden culture and exhibition. As a Standard it takes some years before it comes to profit. F.C.R.H.S. Sept. to Jan.

**RED VICTORIA** (K). A most beautiful culinary fruit which only needs to be seen once to be appreciated. The fruit is large and slightly ribbed and of a bright scarlet colour. Aug. and Sept.

**\*REV. W. WILKS** (K). A very large, culinary Apple of fine form; a good grower and free bearer. Sept. to Nov.

**RIBSTON PIPPIN** (D). Medium; a well-known sort, succeeds best on Paradise stock. The finest fruit is produced on Cordons; not suitable for orchards, should only be planted in warm soils. Nov. to March.

**SIR JOHN THORNEYCROFT** (D). A handsome Apple of good size, round, bright yellow, flushed with red, excellent flavour and good healthy grower. Dec. and Jan.

**ST. EVERARD** (D). Seedling from Cox's Orange, crossed with the old favourite Margil, and the parentage alone is sufficient guarantee of its merit. The fruit resembles Cox's, but more colour. Sept.

PEASGOOD'S NONSUCH.

*If desired we can supply permanent "Acme" Labels for all Fruit Trees. For prices see page 64.*

## Apples (*continued*).

*STIRLING CASTLE (K). An early and free-bearing Apple; a great bearer, and well-suited for dwarf culture. Oct. and Nov.

THE QUEEN (K). A useful culinary apple, free bearer, flesh white and tender. Oct. to Dec.

*VICAR OF BEIGHTON (K). Very handsome deep bright crimson colour, mottled and striped with yellow. Dec. to May.

*WARNER'S KING (K). A very large and splendid Apple of first-rate quality; the tree is a free and vigorous grower, a great bearer, and not subject to disease. Oct. to Dec.

WORCESTER PEARMAIN (K.D.). Handsome early variety, suitable for kitchen or dessert; a great favourite in the market. Sept. and Oct.

## CULTIVATION.

When practicable, the land intended for Apple Trees should be deeply worked all over, but if this cannot be done, a hole 3 ft. square and 2 to 2½ feet deep should be prepared for the reception of each tree, care being taken to break up the hard bottom and to prevent any of the bad subsoil being brought to the surface, or in contact with the roots. The undersoil if bad should be taken out altogether, and replaced by any good sweet loam that is not too rich. When planting, even where the soil has been carefully worked over, it is advisable to make a hole for each tree somewhat larger than the spread of the roots, then throw back into the hole some of the soil so as to form a mound in the centre about as high as the surface of the ground. The tree is then placed on the top of the mound, the roots carefully spread out, and a little soil worked in among them by hand before the bulk of the soil is filled in. Care should be taken that the uppermost roots are not more than about 3 inches below the surface. The soil around the newly-planted tree should then be made firm, and if necessary staked, and a dressing of farmyard manure subsequently applied as a mulch.

When a tree grows too much wood and in consequence bears indifferently, it should either be lifted and replanted forthwith, or else root-pruned.

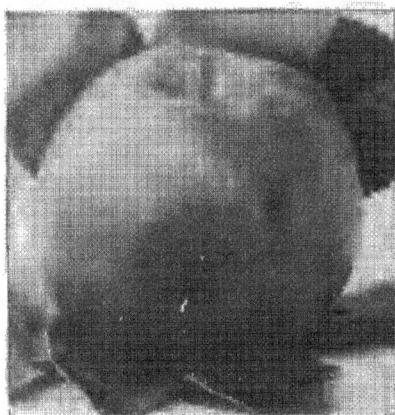

EARLY VICTORIAN.

## CRABS (Pyrus baccata).

The following varieties, which we consider by far the best and most useful, are excellent for making preserves. They are also very pretty as ornamental trees, the bright-coloured fruits hanging in abundance, as they generally do, for a long time in Autumn, being very handsome and effective amongst other ornamental trees, shrubs, &c.

DARTMOUTH. Very handsome dark crimson fruit; an abundant bearer. Standards, 6/6 each.

JOHN DOWNIE. Bright crimson, conical fruit of good size and quality; very handsome. Standards, 6/6 each.

## PEARS.

Pears should be much more freely grown than they are. The young trees come into bearing much earlier than is generally supposed, especially when worked on the Quince stock. Many of the varieties are exceedingly prolific, whilst the fruit are more valuable than Apples, choice sorts always finding a ready sale at good prices.

Our Pears are mostly worked on the ordinary pear stock. We can, however, supply in dwarfs and pyramids on the quince. These come into bearing and profit much earlier than those worked on the ordinary stock. Specially recommended to allotment holders or where tenure is uncertain.

## PRICES OF PEARS ON QUINCE OR PEAR STOCK.

| | | | |
|---|---|---:|---:|
| BUSHES .. .. .. .. .. .. | | 4/- each | 42/- doz. |
| "    Extra strong .. .. .. .. | | 5/- ,, | 54/- ,, |
| PYRAMIDS. Selected .. .. .. .. | | 7/6 ,, | 84/- ,, |
| CORDONS. Single .. .. .. .. | | 5/- ,, | 54/- ,, |
| "    Single, strong fruiting .. .. | | 6/- ,, | 66/- ,, |
| STANDARDS .. .. .. .. .. | | 7 6 ,, | 84/- ,, |

Those marked with an * can be supplied as Standards.

*BEURRE D'AMANLIS (D). Fruit large. Flesh greenish white, fine grained, tender, juicy, melting, rich, sugary and agreeably perfumed. One of the best early pears for market. Sept.

*BEURRE DIEL (D).' Hardy and vigorous, first-rate quality. Flesh yellowish white, tender, a rich sugary flavour. Oct. and Nov.

*BEURRE HARDY (D). A fine, large dessert pear of most excellent quality. As a pyramid it is a very great bearer. Oct.

BEURRE RANCE (D). Large, one of the late pears, of sweet, juicy flavour. Must be root pruned to induce fertility and prevent the fruit cracking; requires a south or west wall.

*If desired we can supply permanent "Acme" Labels for all Fruit Trees. For prices see page 64.*

## Pears—*continued.*

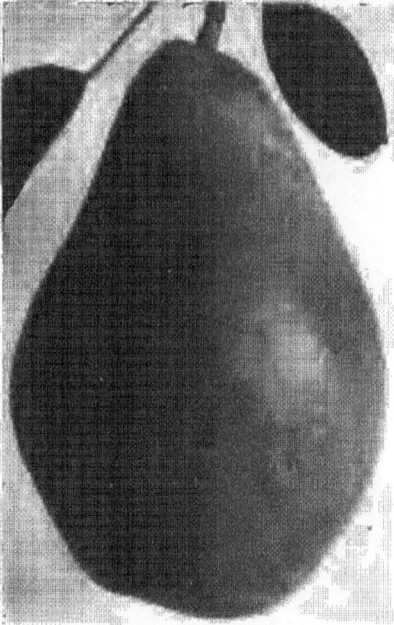

**BEURRE SUPERFIN** (D). One of the best pears in cultivation ; fruit large, very handsome, and of splendid quality. Sept. and Oct.

**BROWN BEURRE** (D). An old favourite and dessert pear. Sept. and Oct.

**\*CATILLAC** (K). Fruit large ; one of the best stewing pears ; does not succeed well as a pyramid or standard unless well sheltered, but is first-class for a wall. Dec. to April.

**°CLAPP'S FAVOURITE** (D). One of the finest early. Medium size, rich yellow colour, flushed with crimson; juicy, delicious flavour. Aug.

**COMTE DE LAMY** (D). Medium, flesh deliciously buttery. Hardy and prolific. Oct.

**CONFERENCE** (D). Fruit large ; skin dark green and russet ; flesh salmon coloured ; melting, juicy, and rich. Makes a strong healthy growth both on the pear and quince stocks ; very prolific. F.C.C. Nov.

**DOYENNE BUSSOCH** (D). A handsome pear of good quality. Oct.

**°DOYENNE D'ETE** (D). Pale yellow with brownish-red flush. Flesh melting, very juicy, white, sweet. July and Aug.

**\*DOYENNE DU COMICE** (D). Large, very handsome, ranks as the most delicious melting pear grown ; it bears remarkably large fruit, and makes a fertile Pyramid, Cordon, or Wall tree. Oct. and Nov.

**DUCHESSE D'ANGOULEME** (D). Of great excellence. Large, greenish yellow, flesh white, buttery and melting, of a rich flavour. Abundant bearer and vigorous grower. Oct. and Nov.

**\*EMILE D'HEYST** (D). An excellent variety of Marie Louise type but hardier, and a more certain cropper. Oct. and Nov.

**\*GLOU MORCEAU** (K). Fruit large, richly flavoured and juicy ; a very fine dessert pear. Dec. and Jan.

**JARGONELLE** (D). Medium. Succeeds on walls or as an open Standard ; makes a prolific tree on the Quince and forms a spreading bush. Aug.

**°JOSEPHINE DE MALINES** (D). A fine pear of most delicious flavour ; the tree is hardy and an excellent bearer. Jan. to May.

**°LOUISE BONNE OF JERSEY** (D). Medium size, skin smooth and shining ; colour pale green with deep chocolate crimson, flesh white, crisp and juicy, very sweet. Oct.

**°MARIE LOUISE** (D). Large to medium, one of the best for walls ; requires frequent root pruning, as it is a rapid grower ; richly flavoured, uncertain in bearing. Oct. and Nov.

**MARIE LOUISE D'UCCLE** (D). Large useful pear of first-rate quality ; great cropper. Oct.

**\*MARGUERITE MARILLAT** (D). Very large and showy, with aromatic flavour ; handsome ; the finest in its season. Sept.

**\*PITMASTON DUCHESS** (D). Very large, melting ; bearing freely on Pear or Quince. Cultivated for market on a large scale, and succeeds admirably as Standards. Oct. to Dec.

**\*ROOSEVELT** (D). A pear of immense size, yellow, tinted with salmon pink, flesh white, melting, and juicy ; a good grower and free bearer on Quince. Oct. to Dec.

**SANTA CLAUS** (D). A splendid pear of large size and handsome appearance which ripens at Christmas ; delicious flavour. Dec. and Jan.

**\*UVEDALE'S ST. GERMAIN** (K). A very large pear, first-class for stewing. Flesh white, crisp, juicy, and slightly gritty. Jan. to April.

**°WILLIAMS' BON CHRETIEN** (D). Well-known splendid old dessert pear ; very hardy and a good bearer. Aug. and Sept.

**WINTER NELIS** (D). Medium, roundish, melting, and deliciously flavoured ; one of the best late pears. Nov. to Feb.

**WINTER ORANGE** (K). A large stewing pear, yellow covered with russet brown ; a good grower and bearer. Feb. and March.

For prices of Bush, Pyramid and Cordon Pears, *see preceding page.*

# HORIZONTAL TRAINED PEARS.

We can supply the following varieties in Horizontal Trained Trees, Two Tiers :—

BEURRE D'AMANLIS (D).
BEURRE DIEL (D).
BEURRE HARDY (D).
CATILLAC (K).
CLAPP'S FAVOURITE (D).
CONFERENCE (D).
DOYENNE D'ETE (D).
DOYENNE DU COMICE (D).
EMILE D'HEYST (D).
CLOU MORCEAU (K).

JARGONELLE (D)
JOSEPHINE DE MALINES (D).
LOUISE BONNE OF JERSEY (D).
MARIE LOUISE (D).
MARGUERITE MARILLAT (D).
PITMASTON DUCHESS (D).
ROOSEVELT (D).
SANTA CLAUS (D).
WILLIAMS' BON CHRETIEN (D).
WINTER ORANGE (K).

2 Tiers, 7s. 6d. each.

# PLUMS.

Although as a rule, Plums attain the greatest perfection when grown as wall fruit, most of the varieties will succeed admirably when grown as Pyramids or Standards. Dwarf trees should be lifted and root-pruned from time to time, if making too strong a growth, and the branches should be thinned out to admit air and induce the formation of fruit buds. For growing on walls the following are highly recommended:—Transparent Gage, Green Gage, Golden Gage, Coe's Golden Drop. For general purposes the following are amongst the very best:—Cox's Emperor, Czar, Early Prolific, Diamond, Jefferson's, Magnum Bonum, Victoria, Monarch, Belle de Louvain, Pond's Seedling.

BUSHES, 5/- each, 54/- per doz.
,, Extra Strong, 6/- each, 66/- doz.
DWARF FAN TRAINED, 7/6 and 10/6 each.

HALF & THREE-QUARTER STANDARDS, 5/- each, 54/- per doz.
STANDARDS, TRAINED Our selection only. 12/6 & 15/- each.

## General List.

*D denotes dessert.    K kitchen.*

**BELLE DE LOUVAIN** (K). Very large handsome fruit of rich flavour. The fruit is red, and is a constant bearer.

**BRYANSTON GAGE** (D). Large, sugary, and richly flavoured; an excellent and abundant bearing sort. Aug.

**COE'S GOLDEN DROP** (D). Large oval fruit, pale yellow spotted with red; one of the very best plums for dessert. End of Sept.

**COE'S VIOLET.** A most valuable variety, delicious flavour, will keep for some time after gathering; skin streaked with yellow and violet, with a very heavy bloom.

**COX'S EMPEROR** (K). Large dark reddish-purple fruit, firm flesh, sweet, rich, and juicy; a very fine bearer. Sept.

**DAMSON, FARLEIGH PROLIFIC.** A splendid market variety. Sept.
,, **KING OF DAMSONS.** A very fine variety of sweet brisk flavour; one of the best.
,, **MERRYWEATHER'S.** Unlike other Damsons it commences to fruit as soon as the trees are two or three years old, after which it never fails. 7/6 each.

**DENNISTON'S SUPERB** (D). Large oval fruit, greenish-yellow blotched purple; a delicious dessert plum, and an abundant bearer. Aug.

**DIAMOND** (K). Large oval, dark purple fruit; excellent for cooking or preserving. Sept.

**EARLY PROLIFIC** (D). Hardy, and a certain bearer; very valuable market sort; ripens middle of July on a wall; one of the best flavoured. July.

**EGG PLUM.** This is the real "Evesham" variety, obovate yellow with golden tinge; a very useful cooking variety, a great bearer.

**ELIZABETH** (K). Very deep purple, bears prolifically. August.

**GOLDEN GAGE** (D). Large fruit, of very rich and delicious flavour; a most excellent and prolific sort. Sept.

**GREENGAGE** (D). The very best and richest of all. This race requires vigorous root pruning, and then bears freely; as Pot trees they succeed in an orchard house. Aug.

**JEFFERSON'S** (D). A large and delicious plum, hardy, good hoarer. Sept.

**JULY GREENGAGE** (D). Medium-sized fruit, yellow, blotched with crimson; an excellent early sort, and a good cropper. July.

**KIRKE'S** (D). One of the very best of the blue plums, the fruit is medium-sized and richly flavoured; a first-rate dessert variety. Sept.

**MAGNUM BONUM, WHITE** (K). Large yellow fruit; an excellent kitchen variety. Sept.

**MONARCH** (K). One of the best late plums, of large size and splendid flavour, dark purplish blue; a heavy bearer and strong grower. Sept.

**ORLEANS** (K). A good cooking or preserving plum; a great bearer. Aug.

**POND'S SEEDLING** (K). Very large, good bearer and a sturdy grower. Forms a spreading tree; valuable for late market or garden culture. Early in Sept.

**PRESIDENT** (new). Very large, late purple, good and free bearer, deep purple with good bloom.

**PRINCE ENGLEBERT** (K.D.). Very large deep purple fruit, sweet and rich and with a brisk flavour, fine for cooking or dessert.

**REINE CLAUDE DE BAVAY** (D). Large, round, greenish-yellow fruit of the "gage" type, rich and delicious flavour; the tree is hardy and a great bearer. Beginning of Oct.

**SHEPHERD'S BULLACE.** Large greenish round fruit, splendid for tarts.

**THE CZAR** (D.K.). Very large, purple fruit of rich flavour; an abundant bearer, most valuable to market growers on account of its earliness, fine appearance, and excellent quality. End of July.

**TRANSPARENT GAGE** (D). Large oval, greenish-yellow fruit marked with red; flesh rich, sweet, and juicy; superior to almost all other plums. Middle of Sept.

**VICTORIA** (K). A well-known and very fine variety; the tree is hardy and an almost constant bearer. The most useful kind of the season. Early in Sept.

**WHITE BULLACE** (K). A very prolific and useful culinary sort. End of Oct.

We can supply all the above varieties in Fan Trained Trees except Diamond and Coe's Violet.

## STANDARD PLUMS.

We are only able to offer the following varieties in STANDARDS, each 7/6, per doz. 84/-.

| | | | | |
|---|---|---|---|---|
| Belle de Louvain | Cox's Emperor | Diamond | Egg Plum | The Czar |
| Bryanston Gage | Cox's Golden Drop | Early Rivers | Prince Englebert | Victoria |

# PEACHES.

DWARF TRAINED, 10/- & 15/- ea. STANDARD TRAINED, 3½ to 5ft. stems(marked*), 25/- & 30/- ea.

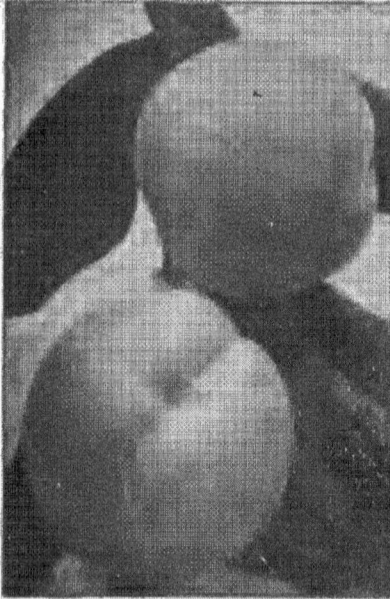

PEACH—ROYAL GEORGE.

*ROYAL GEORGE. Large, very pale, speckled and marbled red, very juicy, rich and highly flavoured; a good bearer. Sept.

ALEXANDRA NOBLESSE. Very large; flesh tender, juicy, and rich. Middle of August.
ALEXANDER. Medium sized, brilliant colour, early peach; skin yellow, almost scarlet next the sun; flesh palo yellow, very juicy. July.
BARRINGTON. Large fruit of rich vinous flavour, and first-rate quality. Sept.
BELLEGARDE. Large, handsome, deep red, almost black next the sun; flesh pale yellow, very juicy. Middle of Sept.
CRAWFORD'S EARLY. Very large, of splendid colour; very tender and melting. Aug. & Sep.
CRIMSON GALANDE. Large; deep crimson; flesh tender, melting, and rich. Aug.
*DR. HOGG. Large, fruit remarkable for its high colour; it is firm yet melting, and of rich sugary flavour. Middle of Aug.
*DUKE OF YORK. Large free stone, fruit well coloured, of excellent flavour; ripe about same time as Alexandra.
DYMOND. Fruit large, skin greenish yellow; flesh white, rich, melting, juicy. Middle of Sep.
*EARLY RIVERS. Large, pale straw-coloured fruit, very rich and fine flavour. End of July.
GOSHAWK. Large, of exquisite flavour, good bearer; pale with red flesh, very hardy. Sept.
GROSSE MIGNONNE. Large, pale greenish yellow, mottled red, deep brown red next the sun; flesh melting, very juicy, of delicious vinous flavour. Early in Sept.
HALE'S EARLY. Medium size, beautifully suffused crimson, flesh melting, juicy and delicious; forces well. July.
*NOBLESSE. Large and handsome, remarkably juicy, with very tender, delicate flesh, sweet and luscious. Sept.
PEREGRINE. Distinct Mid-season variety of fine constitution, large handsome fruit, brilliant crimson skin, flesh rich and highly flavoured.
*PRINCESS OF WALES. One of the largest Peaches and best; skin cream with a rosy cheek, melting and rich. End of Sept.
*SEA EAGLE. Very large, good flavour; remarkable for colour and size. End of Sept.

# NECTARINES.

DWARF TRAINED, 10/- and 15/- each.
STANDARD TRAINED, 3½ to 5 ft. stems (marked*), 25/- and 30/- each.

*EARLY RIVERS. A Seedling Nectarine, raised by Mr. T. F. Rivers, ripening twenty-one days before Lord Napier. It is a certain and heavy cropper, and promises to be one of the most valuable Nectarines yet introduced.
*DOWNTON. Fruit large, oval, skin greenish in the shade, dark red on sunny side; melting, juicy, rich, and highly flavoured; an excellent variety. End of Aug.
*ELRUGE. Medium, pale green, flushed deep red, flesh melting, rich, and juicy; one of the best; an excellent bearer and forces well. Aug. and Sept.
*LORD NAPIER. Medium size; pale cream with red cheek, flesh melting, very early, one of the best.
*PINEAPPLE. Large, bright red on the sunny side, very rich and sweet. Sept.
*PITMASTON ORANGE. Large bright orange, dark brownish-red on the sunny side; melting, juicy, and rich; an excellent Nectarine, a good bearer. Aug. and Sept.
RIVERS' ORANGE. Similar to Pitmaston Orange, but earlier.
VICTORIA. Very large, tender, sweet, and of exquisite flavour; the best late Nectarine under glass.

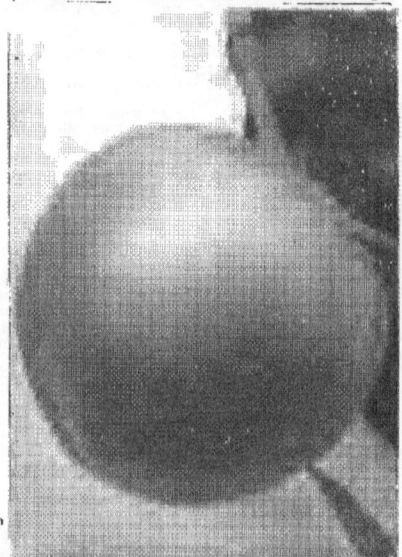

# APRICOTS.

### DWARF FAN TRAINED, 10/- and 15/- each.

The Apricot delights in abundance of fresh air and plenty of light. Like the Cherry, provided the aspect be favourable, it can hardly have too open an exposure. A damp atmosphere and a sour soil on the other hand are its greatest enemies. So susceptible is it to sourness in the soil that wherever limestone is not naturally present a heavy dressing of lime should always be applied before an Apricot tree is planted. On a strong clay soil it is necessary to make a border, excavating the ground to a depth of about 3 feet, and secure effective draining before refilling. At the bottom of the trench it should have 6 inches of brick rubbish, &c., overlaid with chalk or old mortar, the remainder being filled up with a good sound loam, freely intermixed with chalk or lime. No manure should be mixed with the compost, as it is liable to cause sourness.

**HEMSKERK.** Flesh tender, juicy, and richly flavoured. July and Aug.

**KAISHA.** Middle size, flesh deep orange, juicy and rich. Aug.

**MOORPARK.** One of the best. Aug. and Sept.

**PEACH.** Very large, rich, and juicy; one of the finest of all. Aug. and Sept.

**ROYAL.** Large, rich and juicy. July and Aug.

# CHERRIES.

**DWARF FAN TRAINED.** Extra fine trees. 7/6 and 10/6 each.    **STANDARDS.** 7/6 each.
**PYRAMIDS.** 7/6 each.

Cherries thrive on almost any free working, deep, sweet, well-drained soil, provided they have plenty of fresh air. Wherever the soil shows the slightest tendency to sourness, this should be checked by the application of lime.

All Cherries grow well as a rule upon an East wall, Where early crops are wanted a South wall is of course preferable. On a West wall, particularly in a wet district, the fruit is liable to crack. Those marked with an * can be supplied as Standards.

## GENERAL LIST.

*BIGARREAU. Large and of first-rate quality; a capital bearer. July.

*BIGARREAU NAPOLEON. Good bearer, hardy and excellent, follows the Bigarreau; valuable as extending the season; first-rate for market.

BLACK EAGLE. Fruit of good size and flavour; an excellent black cherry. July.

BLACK HEART. A capital early black cherry of good quality. free bearer.

*EARLY RIVERS. Large, shining black, very handsome rich flavour; one of the best for forcing or cherry house, and valuable for wall.

*FROGMORE EARLY PROLIFIC. A capital early sort, very prolific.

*ELTON. Large, rich and excellent. July.

GOVERNOR WOOD. Large, yellow, mottled with red, sweet and rich; a good bearer; excellent. July.

MAY DUKE. Large, juicy, rich, and excellent; an abundant bearer as a standard or a bush. July.

MORELLO. Valuable for preserving and bottling. Pyramid trees produce fruit equal to that from a wall. Succeeds on north walls, and is occasionally planted as a Standard.

*THE NOBLE. Very large, flesh firm, of rich flavour. A profuse bearer, and the fruit keeps well after gathering. Quite distinct.

WHITE HEART. Flesh firm, sweet and pleasant flavoured. End of July.

# CURRANTS.

FAY'S PROLIFIC.

## RED CURRANTS.

**FAY'S PROLIFIC.** One of the best Red Currants. The bush is a strong grower, wonderfully prolific, and comes into bearing early. The fruit is large, bright red, and of excellent flavour.

**Good Strong Bushes, 8d. each; 7/- doz.**

**Extra Strong, 1/- each; 10/- doz.**

## WHITE CURRANTS.

**TRANSPARENT WHITE.** The largest and sweetest of White Currants.

**WHITE DUTCH.** A good currant for general purposes; splendid for dessert.

**Good Strong Bushes,
1/- each; 10/- per doz.**

From Mr. F. W. WILSON, Wakefield.
March 31st.
" I beg to thank you for such a splendid clean sturdy lot of Currants as sent; they are the admiration of all."

From Mr. W. DALLY, Fenwell.
September 26th.
" The Currant Bushes you sent me last year were an excellent lot."

From Mr. W. EVANS, Wattstown.
March 27th.
" I am pleased to tell you that the Currant and Gooseberry Bushes which I obtained from you are doing well, not one failing."

# BLACK CURRANTS.

## SPECIAL OFFER OF STRONG FRUITING BUSHES.

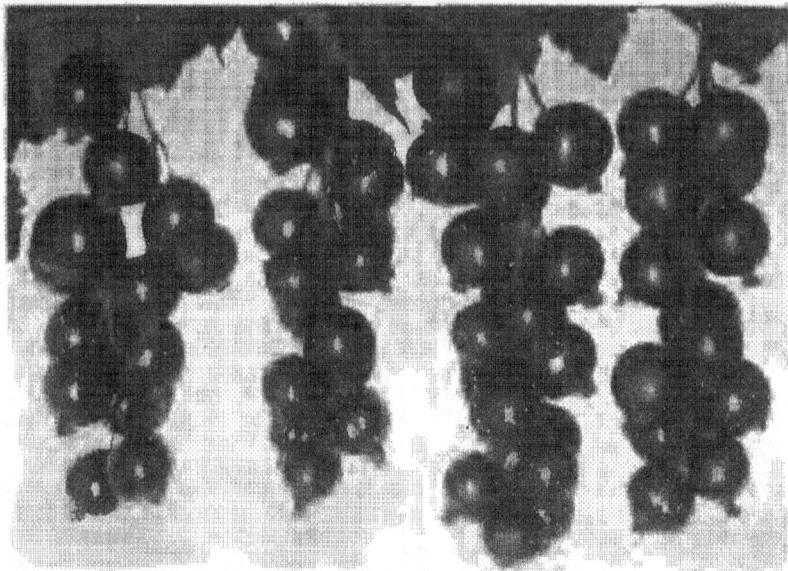

We devote considerable attention to the production of Black Currant Bushes, and our stock is again a very fine one for this Season's Sale. Vigorous in growth whilst sturdy in habit, our bushes may be relied upon to give good crops soon after planting. We propagate our stock only from young clean wood, a most important point to remember, and they are free from disease.

Having a large stock we are able to offer them at lower prices than last year, and to any one having suitable ground it is impossible to devote it to a more remunerative purpose. Black Currants are often thought to be unworthy of much attention—but this is wrong. If carefully planted and regularly mulched with well-rotted manure, the ground being kept clean, there is no garden crop which will give better returns in generous valuable crops.

Orders may be booked for delivery any time after the middle of October, and are executed in date rotation as received.

## BLACK CURRANTS.

**SEABROOK'S BLACK.** A splendid Black Currant which possesses a strong constitution. A most prolific bearer. Our stock is perfectly clean. Those who are troubled with big bud or falling off in fruit are strongly recommended to try it. The best market variety we know.

**BOSKOOP GIANT BLACK.** The finest Black Currant yet introduced. It is of extraordinary vigorous growth with long bunches of enormous fruit. Flavour, sweet and rich. A first-rate variety for exposed situations, and although it flowers late it ripens early.

### PRICES.

| | Each. | Per doz. | Per 100. |
|---|---|---|---|
| Extra Strong Transplanted Two-year old Bushes, Fruiting size, with good roots | 1/- | 10/- | 75 - |
| Good, strong, Two-year old Bushes, well rooted | 8d. | 7/- | 50/- |

Special quotations for large quantities to Fruit Growers and Market Gardeners by return of post.

### NOTES ON CULTURE.

Black Currants thrive best in a deep cool moist soil. On a dry sand or gravel, and on hot shallow soils they are practically useless. They often grow luxuriantly on wet soil, but are liable to disease when the land is sour. A soil containing abundance of humus or vegetable matter suits them well, as such a soil is, as a rule, sufficiently damp for the moisture-loving rootlets, and sufficiently cool to prevent their over stimulation. The surface roots are very sensitive both to mutilation and drought. Ground occupied by Black Currants should therefore be disturbed as little as possible after the bushes are established ; and on the majority of soils a mulching of manure before the advent of the hot season will often preserve the plants from a serious check.

As the finest fruit is produced on last year's wood, pruning must be confined to the removal of old and superfluous wood, the shortening of growths of undue length, and the thinning out or complete removal of suckers, according as the plant is grown as a natural bush or a clean stemmed tree.

# RASPBERRIES.

We have devoted large beds to the production of Raspberries, and our land being especially suitable for them, we are able to offer an exceptionally fine lot of young stuff at moderate prices. We have marked (*) the varieties strongly recommended by us as the result of our trials, and we advise our friends to give preference to these kinds.

Plant eighteen inches apart in the rows, 5 to 6 ft. between the rows.

We have an exceptionally fine lot of the following varieties.

**\*PYNE'S ROYAL.** A grand new Raspberry, the largest in cultivation. It is a strong and upright grower, quite distinct in habit and foliage ; the canes are strong and sturdy and of most robust constitution. The large fruit is borne on short trusses on which they are very thickly set. A limited number of plants only.
1/- each ; 10/- per doz. ; 75/- per 100.

**\*RED CROSS (Pyne).** One of the very best of the new varieties of which we have formed a very high opinion. We have planted out 10,000 on our own Fruit Farm and strongly recommend it. It is a strong grower, quite distinct, the fruit are of large size and borne in the greatest profusion, splendid for either cooking or Jam making.
1/3 each ; 12/6 per doz. ; 95/- per 100.

**\*THE DEVON.** Grand new Raspberry, growth most robust ; trusses have been found to carry as many as sixty fruits ; an enormous cropper, brings all its fruit up to a large size. For Jam making and Bottling it is of the highest value.
1st size, 8d. each ; 6/- doz. ; 45/- 100.
2nd size, 6d. each ; 4/6 doz. ; 32/6 100.

**\*BAUMFORTH'S SEEDLING.** A fine variety ; fruit very large, of the most beautiful crimson colour ; an abundant bearer of good habit.
1st size, 25/- per 100 ; 4/- per doz.
2nd size, 18/- per 100 ; 3/- per doz.

**HAILSHAM.** A new Autumn fruiting Raspberry ; the fruit is very large, of a rich crimson colour and excellent flavour ; strong grower and heavy bearer. 45/- por 100,; 6/- per doz.

**\*HORNET.** A very fine Raspberry ; fruit deliciously flavoured, and the most juicy of any variety. A splendid cropper.
1st size, 25/- per 100 ; 4/- per doz.
2nd size, 18/- per 100 ; 3/- per doz.

**PROFUSION.** A very strong grower and great bearer, fruit round, large, bright red, and of very fine flavour.
1st size, 25/- per 100 ; 4/- per doz.
2nd size, 18/- per 100 ; 3/- per doz.

**\*PERFECTION.** It is an exceptionally strong grower and makes a better plant the first year than any other variety. It is a good cropper, producing fruit from base to top of cane of a bold size, firm, fine, aciduous flavour, and brilliant scarlet colour, the high colour being retained even when the fruit is fully ripe.
1st size, 25/- per 100 ; 4/- per doz.
2nd size, 18/- per 100 ; 3/- per doz.

**SUPERLATIVE.** Fruit very large, mostly freely produced ; au excellent variety.
1st size, 25/- per 100 ; 4/- per doz.
2nd size, 18/- per 100 ; 3/- per doz.

### CULTIVATION

Ground intended for these should be deeply trenched and heavily manured. The canes should be planted (not too deeply) about 12 to 18 inches apart, and the rows should be 5 or 6 ft. apart, and after planting, a mulching of well-decayed manure should be placed on the surface. Newly planted canes should be cut back to 2 feet to encourage the formation of suckers for the following season.

## LOGAN BERRY.

This fine American fruit has proved itself to be a decided acquisition, being hardy and very productive, and is extensively grown for market purposes. Grown in the same way as Raspberries it fruits freely, and will often succeed where these fail. It is of very strong growth, and makes a capital plant for poles or pillars. The fruit is large and of a deep rich red colour, possessed of a rich luscious flavour, and is well suited for dessert or culinary purposes. We have an exceedingly fine lot of strong plants from layers, the true variety. 1/6 each, 16/- per doz., and 2/- each and 21/- per doz.

From **R. E, HOLIDAY,** Catford, S.E.
Jan. 29th.
" The **Logan Berries,** two plants, which I had from you two or three seasons ago, have borne splendidly. We made about 50 lbs. of Jam from same beside using a large quantity for stewed fruit. I also took first prize at our local show with them."

From **MRS. WOODS,** Malvern.
July 11th.
" The **Raspberry** canes which I purchased from you in the Spring have done very well, considering the short while they have been planted."

# GOOSEBERRIES.

GOOSEBERRY, WHINHAM'S "INDUSTRY."

There has in recent years been a great demand for Gooseberries, and we have extended our culture of them considerably.

The Gooseberry, like many other fruits, prefers plenty of fresh air and a sunny position, except in hot dry localities. Though a shallow rooting plant, the need of moisture makes it prefer a soil containing abundance of humus. Strong clay and light sands can both be greatly improved in condition for this crop by the addition of plenty of farmyard manure, leaf mould, and vegetable matter generally.

## LANCASHIRE PRIZE VARIETIES.

A very fine class, much esteemed for the splendid size of their fruit and their value for exhibition or dessert. When well ripened they are of delicious flavour and equal to many forced fruits.

WHITE.    RED.    YELLOW.    GREEN.

Our selection only.

**Bushes, 1/6 each, 16/- per doz.**

CROWN BOB. Large red fruit with very thin skin and of the finest flavour, can be picked early in a green state.

GOLDEN DROP. Golden yellow, medium size, very early, delicious flavour.

KEEPSAKE. A very large straw-coloured variety of excellent flavour, and one of the best and earliest for gathering green.

WHINHAM'S "INDUSTRY." A superb variety, bearing a wonderful profusion of large handsome fruit, which are of a dull red colour when ripe.

WHITESMITH. White, large fruit, splendid flavour, an exceedingly heavy cropper; early.

**Bushes, 1/- each, 10/6 per doz.**

Standard Gooseberries on 2 to 3 ft. stems, our selection. 5/- and 6/6 each.

## ALMONDS.

Two-year Rods 6/- each.

## BLACKBERRIES.

Most of them are quite hardy, and succeed well under similar culture to the Raspberry. The fruits are large, handsome and delicious, either raw, cooked, or preserved.

PARSLEY LEAVED. Very ornamental cut-leaved variety, which bears large fruit, good and productive. 1/- each ; 9/- per doz.

WILSON JUNIOR. Large fruit; delicious. 1/- each ; 9/- per doz.

## MEDLARS.

For the successful cultivation of Medlars an open situation sheltered from cutting winds is absolutely essential. A good moist well-drained loam suits them best; but with an occasional mulching they grow well on sandy soils.

Standards 6/6 and 7/6 each.

## QUINCE.

Standards 7/6 each.

## WALNUTS.

Fine Standards 10/6 and 12/6 each.

## RHUBARB.

CHAMPAGNE. Deep red, early. 2/- each ; 21/- per doz.

DAWS' CHALLENGE. A remarkable variety for forcing, growing up to 4 ft. Very fine colour, First Class Certificate, R.H.S. 3/6 each.

DAWS' CHAMPION. Very large. 2/6 each; 28/- per doz.

VICTORIA. The stalks are large and of good quality, a well-known and excellent variety for summer use. 2/6 each.

# SPECIAL
# COLLECTION of BUSH FRUIT TREES.

This offer is made to meet the needs of those having small gardens or allotments, and who have not room for a large number of Fruit Trees.

| | | |
|---|---|---|
| 4 RED CURRANTS. | 12 RASPBERRIES. | 8 BLACK CURRANTS. |
| 2 APPLES, Bush. | 6 GOOSEBERRIES. | 2 PLUMS, Bush. |
| | 1 BUSH PEAR. | |

The above 35 Bushes for 35/- Packing Free, Carriage Extra.

We always have a large demand for this Collection. The selection of Varieties must be left entirely to ourselves, and no alterations can be made in any way whatever.

# GRAPE VINES.

Many good crops of grapes are grown in greenhouses without any artificial heat, and often when other plants are cultivated in the same house. Great care must however be taken that plants are not grown which are liable to be attacked by Mealy Bug or Red Spider. Ferns, Palms, Bulbs, Chrysanthemums and Zonal Pelargoniums may however be grown in a vinery. The best position for a vinery is facing South or South-West, and a lean-to house is the best.

In making the border for planting, it is most important that ample drainage be supplied, especially so if the site is at all cold or wet.

To those about to plant vines we recommend the careful study of some good book, such as "GRAPES AND HOW TO GROW THEM," by J. Landsell, price 3/9 post free.

The fruiting canes we offer are strong and stout, from eight to ten feet in length; and if cultivated in pots will bear from eight to twelve bunches each next season.

*H denotes those varieties that require a heated vinery. C denotes those suitable for growing in a cool vinery*

**BLACK ALICANTE (H).** One of the largest and best grapes for late work, carrying a fine bloom.

**BLACK HAMBURG (C).** Juicy, sweet and rich; a well-known and excellent sort, sometimes ripens out of doors; the best for general use, pot culture, and forcing.

**FOSTER'S SEEDLING (C).** One of the finest and most easily cultivated of White Grapes; and a certain cropper.

**GROS COLMAR (H).** Berries very large, round, jet black with a beautiful bloom. Late and hangs well.

**LADY DOWNES' SEEDLING (H).** A first-rate black late hanging Grape of excellent flavour.

**MADRESFIELD COURT.** A very handsome black Muscat, with large oval berries, covered with a dense bluish plum-like bloom, branches very long and tapering. An excellent variety for early use.

**MUSCAT OF ALEXANDRIA (H).** Rich amber; bunch and berries immensely large, with a deliciously rich, sweet Muscat flavour; requires a warm vinery; the favourite "Muscat."

**ROYAL MUSCADINE.** A fine hardy white; succeeds well on a south wall.

**SWEETWATER BUCKLAND (C).** A round white early Grape, very showy and handsome.

**STRONG PLANTING CANES.**
In pots, 10/6 each.
**FRUITING CANES.**
In pots, very fine, 12/6 and 15/- each.

# FIGS.

Figs will grow in almost any soil, but if it be too rich they produce a great deal of wood and very little fruit. Exuberance of growth is one of their chief characteristics. This can be restrained by limiting their rooting area and making the soil firm. They require a considerable amount of moisture when the fruit is swelling. Good drainage is also essential.

We have an extra fine lot of trees of the varieties named below.

**STRONG PLANTS, in pots, Trained flat.**
**FRUITING　,,　　,,**
**BROWN TURKEY.** Most abundant bearer; the finest for out door.

**BROWN TURKEY. Strong Plants,** Fan-trained, from open ground. 12/6 each.
**WHITE ISCHIA.** Small, sweet and delicious; produces three crops a year in heat; forces well; great bearer; for indoor culture only.

STRONG PLANTS, 5/-, 7/6 & 10/6 each.

# NUTS AND FILBERTS.

Nuts should be planted by preference on the highest and driest available ground. As they are often injured by frosts at the flowering season, it would be well if advantage were taken of the shelter provided by adjacent trees, to protect them in a measure from the cutting winds of Spring. They grow well in stony land, provided it be thoroughly well-drained, and that there be a fair admixture of soil. Suckers should be removed as soon as they appear, as they greatly interfere with the fruitfulness.

We have a very fine stock of these in good strong plants, comprising such varieties as Cosford, Kentish Cob, Filbert, white, red, purple-leaved, &c.

**STRONG FRUITING DWARFS or BUSHES in first-class condition.** 2-3 ft. 2/- each; 21/- per doz.
,,　　,,　　,,　　,, Extra strong 3-4 ft. 2/6 ,, 27/- ,,

# STRAWBERRY PLANTS.

## 1922 Prepared Runners. For Delivery in September and October.

PACKING AND POSTAGE—On 100 Runners, 1s. 3d. ; 50 Runners, 1s. ; 12 Runners, 6d.

Strawberries should be planted as early in the Autumn as possible, Ground that has been heavily manured for early potatoes or Spring cabbages is usually in good condition for the crop. Land that has not been recently manured or thoroughly worked over should be deeply trenched and plenty of farmyard manure added during the operation. When the bed is ready for planting the soil should be levelled and rolled or trodden, unless it is naturally very strong in texture. Planting should be done with a small spade or trowel. The roots of the plants should be carefully spread out and the soil above them firmly pressed with the foot, taking care that the crown is just above ground, and that there is no depression round the neck of the plants.

**KENTISH FAVOURITE.** The heaviest cropping Strawberry yet sent out. The fruit is of a beautiful bright scarlet colour, rather flattish in shape, and superior in flavour to Royal Sovereign. 10′- per 100, 5/- per 50, 2/- per doz.

**KING GEORGE.** A grand new early variety, equalling Royal Sovereign in size, but ripening fully a week earlier. 10,- per 100, 5/- per 50, 2/- per doz.

**LAXTONIAN.** This new variety is the best maincrop yet introduced ; the fruit is large, the centre ones wedge-shaped, the trusses are strong and bold. The flavour is first-class, and we do not know any other Strawberry equal in size so good. 15/- per 100, 7/6 per 50, 3/- per doz.

**LAXTON'S MAINCROP.** The following is the raiser's description :—" The largest and firmest Strawberry we have yet raised, a cross between Bedford Champion and The Laxton ; a vigorous grower, throwing out very bold large trusses in the greatest profusion." 10/- per 100, 5/- per 50, 2/- per doz.

**PRESIDENT.** Great cropper, colour crimson, of superior flavour. 10′- per 100, 5′- per 50, 2/- per doz.

**ROYAL SOVEREIGN.** Fine early variety. The best for pot culture. 10 - per 100, 5/- per 50, 2/- per doz.

**SIR JOSEPH PAXTON.** Hardy early variety. The sweetest and most reliable Strawberry grown. 10/- per 100, 5/-per 50, 2/- doz.

**ST. ANTOINE DE PADOUE.** Perpetual fruiting variety. 10/- per 100, 5′- per 50, 2/- per doz.

**THE DUKE.** A really good early variety ; it forces well, fruit oval, medium size, good flavour. 20/- per 100, 10/- per 50, 4′- doz.

**BEDFORD CHAMPION.** One of the largest fruits in commerce, 2½ to 3 ozs. in weight. Bright scarlet skin, flesh white. 10/- per 100, 5/- per 50, 2/- per doz.

**GIVON'S LATE PROLIFIC.** Large, wedge-shaped fruit of rich colour, and splendid flavour. Award of Merit, R.H.S. 15/- per 100, 7/6 per 50, 3/- doz.

**HIBBERD'S KING GEORGE.** This is a large late Strawberry. Fruit wedge-shaped, but some come pointed. 10′- per 100, 5/- per 50, 2/- per doz.

### 100 in 4 Choice Varieties, our own selection, 10/- per 100.

---

## SWEET AND POT HERBS.

We have a fine collection of these, including the following sorts :—

| | | |
|---|---|---|
| Balm | Mint, Lamb | Sage, Common |
| Chamomile | Mint, Pepper | Savory, Winter |
| Horehound | Pennyroyal | Thyme, Lemon |
| Lavender | Rosemary | Thyme, Common |
| Marjoram | Rue | Wormwood |

Per doz. 7/6, 9d. each.

---

## SEA KALE.

This valuable esculent is easily forced if care is only taken to apply heat gradually, as it will not succeed if placed in too high a temperature at starting. Place several crowns a few inches apart in large pots, and stand them in a temperature of about forty-five degrees, with an inverted pot placed over each to exclude light and insure blanching ; a mushroom house, pit, or cellar, will do well for this purpose. Sea Kale may also be easily forced in the open ground by covering it over with large specially-made pots and applying fermenting material.

**STRONG PLANTING ROOTS.** 4/- per doz., 27/6 per 100.
**GOOD STRONG ROOTS,** for forcing. 5/- per doz., 35/- per 100.
**EXTRA STRONG ROOTS,** for forcing. 5/6 per doz., 40/- per 100.

# NOTES ON THE CULTURE OF HARDY FRUIT.

For small occupations and allotment gardens Dwarfs or Pyramids are best planted eight to twelve feet apart on each side of the path, and about three feet from the paths.

The best situation for Fruit growing is a fairly open piece of ground, protected from the east and north-east if possible, as the winds from these quarters, when the trees are in blossom in Spring, often are most injurious, and ruin the crop for the whole season. A good deep loam is the ideal soil, and the preparation before planting the trees should receive most careful attention.

If the ground is heavy and cold, it should be thoroughly drained, and receive a good dressing of lime or ashes, which should be thoroughly incorporated with the soil to lighten it ; on the other hand, if a light sandy soil is the only piece available, a dressing of clay or brick earth will give body to it and be of much value.

Probably the cause of failure with most people is to be found in not using care in planting the trees ; too often the bole made for the tree is not large enough, and consequently the roots are crowded together in a bunch, and they cannot thrive as they should ; it is of the utmost importance that the holes should be sufficiently large to allow of the roots being laid straight out, and if this is done and the soil carefully shaken between the roots, being at the same time made gradually firm, and the stem quite secure, little fear need be entertained as to the future.

Where it is desired to plant standard fruit trees in grass land or orchards, it is a good plan to pare off the turf for a space of six feet square, and incorporate the chopped up turf with the soil, which is removed to make the hole. A good layer of decomposed manure should be placed at the bottom of the hole, and when the trees have been planted (as advised above) a stout stake should be placed to each tree, being secured by a hay band or some other material that will gradually give and allow the stem to swell, at the same time keeping the tree from swaying about in the wind.

In planting fruit trees on grass land it is advisable to remember that the grass should not be allowed to grow within three feet of the stems, but the ground should be kept forked, and clear of weeds, thus allowing the air to have access to the soil, and rain to penetrate.

Standard trees should be planted about twenty-four feet apart (not less), thus allowing space for the tree and roots to spread and flourish.

Newly planted trees should receive attention during the first few years, if possible being well watered during the dry season ; should it be impossible to water them a dressing of manure or grass round the stem will do much to retain the moisture by keeping off the hot sun.

Fruit trees making very gross growth should be root pruned by partially lifting the tree and passing a spade underneath to sever the coarser roots.

During recent years much attention has been directed to the spraying of Fruit trees, so as to destroy the caterpillars which work such havoc, both with the growth of the tree, and with the fruit when developed. We have no hesitation in recommending regular spraying with some approved insecticide, both when the trees are dormant, to prevent the moth from secreting itself in the bark, and, also, when the trees are in leaf, and the fruit formed, to rid the trees of the fully developed caterpillar which will rapidly make the crop of fruit worthless.

We shall be glad to give further advice to our customers, free of charge, on the selection and planting of orchards, or fruit trees of any kind, or on the treatment of diseases.

Below we give a diagram showing the various forms in which fruit trees are usually sent out from the Nurseries.

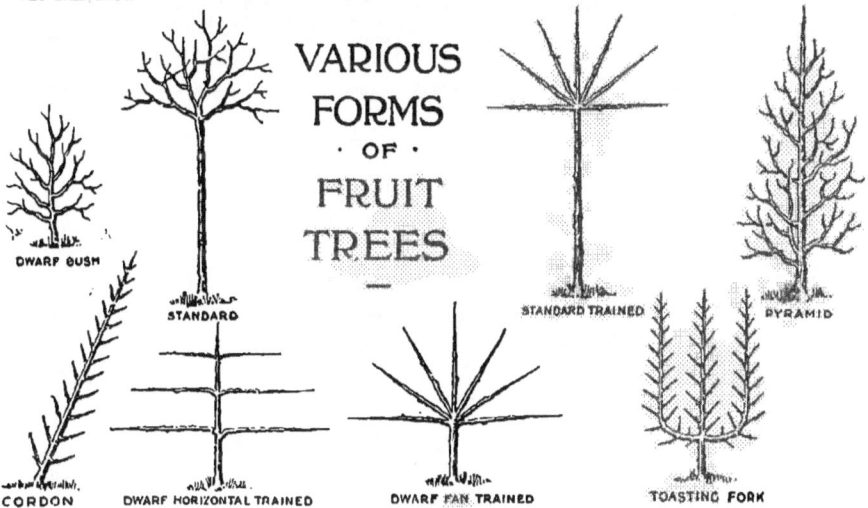

VARIOUS FORMS · OF · FRUIT TREES

DWARF BUSH

STANDARD

STANDARD TRAINED

PYRAMID

CORDON

DWARF HORIZONTAL TRAINED

DWARF FAN TRAINED

TOASTING FORK

From Miss BURTON, Skegness.

May 19th.
" Every second year one of the Pear Trees is simply weighed down with Fruit."

From Mr. ATKINS, Happlowell.

September 26th.
" The Fruit Trees I got from you last year have been fine, especially the Lane's Prince Albert Apples."

# COLLECTIONS
### OF
# CHOICE ROSES
## BRITISH GROWN.

The Hybrid Tea Roses named in the following list are by far the most beautiful and useful for general cultivation in the Garden or for Table decoration or Exhibition. They commence blooming early in June and continue to furnish some fine bloom to the end of October. We have a fine collection of these, in good strong, healthy and well-grown plants, and would also draw attention to our fine stock of Climbing, Pillar, and other Roses.

When special colours are needed for the carrying out of Colour Schemes we ask that our customers will place their orders as soon as possible, as we are invariably sold out of some sorts before the end of the season.

## The "NORWICH" Collection of
### 12 Good BUSHES in 12 extra choice varieties 18/-.
*Packed and Carriage Paid.*

COVENT GARDEN. Rich deep crimson.
DAILY MAIL. Shrimp pink
GOLDEN EMBLEM. Beautiful yellow
GORGEOUS. Deep orange yellow
GEORGE DICKSON. Velvety black crimson
HUGH DICKSON. Brilliant crimson.

LADY PIRRIE. Shrimp apricot
LOS ANGELOS. Flame pink
M. D. HAMILL. Straw colour
MISS WILLMOTT. Cream
MRS. W. QUIN. Canary yellow
OPHELIA. Salmon rose

## The "TOWN CLOSE" Collection of
### 18 Very Choice BUSH ROSES for Exhibition 25/6 *Carriage and Packing Paid.*

CAROLINE TESTOUT. Bright satiny rose.
COLONEL O. FITZGERALD. Red velvety crimson
DUKE OF EDINBURGH. Bright vermillion.
FLORENCE FORRESTER. Clear snow white
FRAU KARL DRUSCHKI. White
GENERAL McARTHUR. Dark velvety scarlet
GEORGE DICKSON. Velvety black crimson
GOLDEN OPHELIA. Golden yellow
HOOSIER BEAUTY. Glowing crimson
JULIET. Old gold

LADY ASHTOWN. Pale rose
LA FRANCE Bright lilac rose
MADAME ABEL CHATENAY. Bright Carmine rose
MADAME RAVARY. Golden yellow
MILDRED GRANT. Ivory white
MISS MAY MARRIOTT. Terra cotta
MRS. A. R. WADDELL. Rosy scarlet
PRINCE CAMILLE DE ROHAN. Velvety crimson

NOTE:—As the planting season advances it may be necessary to alter some of these varieties but we shall send in place other equally good sorts.

## The "POPULAR" Collection of
### 12 BUSH ROSES in 12 good varieties, free blooming, with a good gradation in colour, and sure to give satisfaction.
#### Our Selection 15/- *Carriage and Packing free.*

## The "CLIMBING" Collection of
### 12 free growing beautiful CLIMBERS, including the most popular varieties.
#### Our Selection 18/- *Carriage and Packing free.*

Under this heading we include all varieties suitable for Walls, Arches, Pergolas and Bowers, and if our customers will kindly state, when ordering, for which purpose they are required we shall be very pleased to send varieties to suit their purpose, as far as possible.

## HYBRID SWEET BRIAR.

ANNE OF GIERSTEIN. Dark crimson.
BRENDA. Maiden's blush, peach, dainty colour.
FLORA McIVOR. Pure white, blushed with rose.
LADY PENZANCE. Beautiful soft tint of copper.

LORD PENZANCE. Soft shade of lawn, yellow in centre.
MEG MERRILEES. Gorgeous crimson, very free flowering.
ROSE BRADWARDINE. Beautiful clear rose.

The above varieties 1/6 each; 17/- per doz.

# BRITISH GROWN ROSES.
## NEW AND VERY CHOICE SORTS.

We offer below a selection of New Roses, all the best varieties of recent introduction. The whole of the varieties offered have been tested by us, and they may be relied upon as being the most beautiful in each colour.

To those who are renewing their stock of Roses, and who desire to procure the very best, we strongly recommend this list.

Although we hold good stocks of most sorts, we request that orders be placed as early as possible, as we anticipate a great demand this Autumn.

**SUNSTAR** (H.T.). The most distinct and beautiful rose yet produced, the many phases of colour depicted are almost too intricate to describe, including deep orange and yellow edged, veined and splashed crimson and vermilion, flowering in great profusion continuously throughout the season. Gold Medal N.R.S. **3/6 each.**

Varieties priced at 3/- each, 33/- per doz.

**ASPIRANT MARCEL ROUYER** (H.T.). Long pointed buds, shell shaped flowers, deep apricot tinted salmon flesh, fine for exhibition. **3/- each.**

**BENEDICTE SEGUIN** (H.T.). Reddish apricot bud shaded with carmine, large full globular flower, colour Roman ochre shaded with coppery orange. **2/6 each.**

**EARL HAIG** (H.T.). Deep reddish crimson, solid colour does not fade, immense size and perfectly formed fine Exhibition Rose, particularly good in autumn, sweetly perfumed. **3/6 each.**

**ELIZABETH CULLEN** (H.T.). Intense dark crimson, buds long and pointed, free flowering, exquisitely perfumed, a superb bedding variety. Gold Medal N.R.S. **3/6 each.**

**ETHEL SOMERSET** (H.T.). Shrimp, pink edge of petals deep flesh coral pink, fine shade of colour and very fragrant, a fine rose for exhibition, bedding, and forcing purposes. **3/6 each.**

**GLORY OF STEINFURTH** (H.T.). Cherry and geranium red, very fragrant, splendid for massing. **3/- each.**

**HAWLMARK CRIMSON.** Pointed bud, crimson crayonings on maroon, crimson scarlet as blooms develop, fine decorative and bedding rose. Gold Medal N.R.S. **2/6 each.**

**HORTULANUS BUDDE** (H.T.). Dark red, very free flowering. **2/6 each.**

**LAMIA** (H.T.). Reddish orange, medium size charming bud. Gold Medal N.R.S. **3/- ea.**

**MIRIAM** (H.T.). Nasturtium red, distinct globular flowers carried erect, free flowering and continuous. Gold Medal N.R.S. **3/- ea.**

**PADRE** (H.T.). Upright growth, producing flowers with long petals, coppery scarlet flushed yellow at base of petals. **2/6 each.**

**SOUVENIR DE CLAUDIUS PERNET** (Pernetiana). Striking sunflower yellow, beautifully formed, with elongated deep petals, brilliant green foliage which is immune from all diseases. **3/- each.**

Varieties priced at 3/6 each, 39/- per doz.

# HYBRID PERPETUAL & H.T. ROSES.
## PRICES OF BRITISH GROWN ROSES (Customer's Selection).

Varieties priced at 1/6 each, 17/- per doz., 135/- per 100.
Varieties priced at 2/- each, 21/- per doz., 170/- per 100.
Varieties priced at 2/6 each, 27/- per doz., 215/- per 100.

**ADOLPH KARGER** (Pernetiana). Chrome yellow, large flowers, fine. **2/6 each.**

**AUGUSTUS HARTMANN** (H.T.). Geranium red flushed with red. **2/- each.**

**BETTY** (H.T.). Coppery rose, overspread with golden yellow, perfectly formed flowers. A continuous bloomer. **1/6 each.**

**CAPTAIN GEORGE DESSIRIER** (H.T.). Vigorous grower of spreading habit, large full flower, globular, beautiful dark red shaded with crimson and fiery-red. **2/6 each.**

**CAROLINE TESTOUT** (H.T.). Bright satiny-rose, with brighter centre, large, full, and globular, very free. **1/6 each.**

**CHAMELEON** (H.T.). Outside petals rosy pink, centre shrimp pink, edges of petals fimbriated; a fine rose and will last a long time out. **2/- each.**

**☞ BRITISH QUEEN** (H.T.). The finest pure white Hybrid Tea Rose in existence. **1/6 each.**

**CHATEAU DE CLOS VOUGEOT** (H.T.). Red velvety scarlet, shaded fiery red, passing to dark velvety crimson, which is said to keep its colour under a hot sun. Flowers large, full and globular; a grand variety. **1/6 each.**

**CHEERFUL** (H.T.). The colour is a pure soft orange flame with a distinct orange-yellow base, a combination hitherto unknown in Roses. The blooms are very large and of perfect shape and form, with enormous petals, and is one of our series of mildew-proof roses. Very sweetly scented. **2/- each.**

**CHRISTINE** (H.T.). The deepest and clearest yellow yet seen in Roses, perfectly faultless in shape and form. The colour is so bright that it sparkles with glittering intensity. Mildew proof. Gold Medal. **1/6 each.**

## British Grown Roses (H.P. & H.T.)—General Select List (*continued*).

GORGEOUS.

**CLEVELAND** (H.T.). Coppery yellow at base of petals, heavily flushed reddish copper. 2/- ea.

**COLONEL O. FITZGERALD** (H.T.). Blood red velvety crimson, beautifully finished blooms, produced in great profusion on erect stems, sweetly scented. 2/- each.

**COLUMBIA** (H.T.). Flower true pink, deepening as it opens to glowing pink, produced on stiff stems ; very fragrant. 2/- each.

**CONSTANCE.** Very fine, colour clear cadmium yellow, large exhibition flowers. 1/6 each.

**COUNTESS CLANWILLIAM.** Delicate peach, pink at base of petals, edged with deep cherry red. 1/6 each.

**COVENT GARDEN** (H.T.). Rich deep crimson, well formed flowers on stout stems, branching growth, glossy and mildew proof foliage, very free and late bloomer ; useful for bedding and forcing. 2/- each.

**☛ DAILY MAIL.** Very vigorous, coral red bud shaded with yellow on the base. 1/6 ea.

**DUKE OF EDINBURGH.** Very bright vermilion, a distinct variety. 1/6 each.

**EDITH CAVELL** (H.T.). A splendid variety for exhibition purposes, colour opening white tinged with cream, base of petals pale yellow, long pointed buds, robust grower. 2/- each.

**☛ EDITH PART** (H.T.). Colour rich red with a suffusion of deep salmon and coppery yellow ; fine exhibition rose. 1/6 each.

**EMMA WRIGHT** (H.T.). Pure orange ; an attractive bedding variety. 2/- each.

**FLORENCE FORRESTER.** Clear snow white, with lemon tinge, large, sweet-scented ; the best white for exhibition. 1/6 each.

**FRANKLIN** (H.T.). A fine bedding rose of the type of Lady Roberts, very free flowering, long pointed buds, colour salmon pink, with deeper apricot shadings ; a fine rose for cut flowers and all decorative work. 2/6 each.

**FRAU KARL DRUSCHKI.** (H.P.). Flowers large, perfectly formed with well-shaped petals, and of the purest white, opening well. 1/6 each.

**GENERAL McARTHUR** (H.T.). Dark velvety scarlet ; very sweet-scented. 1/6 each.

**FRIEBURG** (H.T.). Strong growth, fine foliage, and great well-shaped flowers of a lovely rose pink on the outside of the petals, and nearly straw white on the inner side. 2/6 each.

**GEORGE C. WAUD** (H.T.). A brilliant vermilion colour, orange base, with high pointed centre, does not fade. 1/6 each.

**☛ GEORGE DICKSON** (H.T.). Velvety black crimson, the back of the petals being heavily veined with pure crimson. 1/6 each.

**GOLDEN EMBLEM** (new) (H.T.). A great improvement on Rayon d'Or, the colour being deeper and richer, with larger and more perfect blooms. It even surpasses the well-known favourite " Marechal Niel " ; the habit of growth is splendid ; it is mildew proof, very sweetly scented. 1/6 each.

**GOLDEN OPHELIA** (H.T.). A seedling from the well-known and justly admired variety Ophelia. The blooms are borne on a very stout upright stem of a good size, opening in perfect form, golden yellow in centre, paling slightly towards the outer petals. 1/6 each.

**GORGEOUS** (new) (H.T.). Deep orange yellow flushed coppery yellow ; a very beautiful variety. 1/6 each.

**HADLEY** (H.T.). Bright red. 1/6 each.

**HENRIETTE** (H.T.). Fiery orange crimson to soft coral salmon, shaded at base with glowing orange. Strong plants, 1/6 each.

**HOOSIER BEAUTY.** Glowing crimson with darker shading, exceedingly free ; fine show and bedding variety. 1/6 each.

**H. V. MACHIN** (new). Scarlet crimson, very fine, large, full, well-formed flowers, with high pointed centre. 1/6 each.

**HUGH DICKSON** (H.P.). Intense brilliant crimson shaded scarlet. Large and fragrant. 1/6 ea.

**INDEPENDENCE DAY.** Vigorous growth, mildew proof foliage. Buds olive shaped developing into well modelled flowers with flame coloured stains, petals of sunflower gold, rich fragrance. 2/6 each.

**IRISH FIREFLAME.** Deep madder-orange splashed with crimson. 1/6 each.

**ISOBEL** (new). The most beautiful single rose in cultivation. It is carmine red flushed orange-scarlet, with a faint Austrian copper shading, the centre is a pure yellow zone. 1/6 each.

**JEAN G. N. FORESTIER.** Lincoln red passing to carmine lake, slightly tinted with Chinese orange yellow. 1/6 each.

**JULIET.** Outside petals old gold, interior rich rosy red, base of petals deep yellow ; large flowers, distinct. 1/6 each.

**K. OF K.** A startling dazzling semi-single rose of intense scarlet of absolutely pure colour ; its huge petals are velvet sheened, solid scarlet throughout. 2/- each.

**LADY ASHTOWN** (H.T.). Colour very pale rose, shading to yellow at base of petals, reflex of petals silvery pink. 1/6 each.

**LADY PIRRIE** (H.T.). The outside of petals are deep coppery reddish salmon, inside of petals apricot. 1/6 each.

**LA FRANCE** (H.T.). Bright lilac rose ; beautiful. 1/6 each.

**LIBERTY** (H.T.). Brilliant crimson, large, elongated, beautifully formed buds ; a splendid Rose for bedding. 1/6 each.

**LOS ANGELOS** (H.T.). A very fine improvement on Lyon Rose, a much better grower and bloomer, colour flame pink shaded to yellow toned with salmon. 2/- each.

**LOUISE BALDWIN** (H.T.). Rich orange with soft apricot shading over the entire petal, a beautiful even colour, a big advance on all roses of the " Lady Hillingdon " type, a good grower and fine for bedding ; very sweetly scented. 2/- each.

**LOUISE CATHERINE BRESLAU** (H.T.). Coral red, shaded creme yellow, opening to shrimp pink ; very vigorous. 1/6 each.

All our Bushes are staked in the growing season, and they are therefore thoroughly well ripened.

## British Grown Roses (H.P. & H.T.)—General Select List (*continued*.)

RICHMOND.

**LYON ROSE** (H.T.). Edges of petals pink, centre salmon-pink shaded with chrome yellow. 1/6 each.

**MADAME ABEL CHATENAY** (H.T.). Bright carmine rose shaded to deep salmon; long pointed full-sized flowers. 1/6 each.

**MADAME MELAINE SOUPERT** (H.T.). Colour pale saffron yellow, suffused with pink and carmine. 1/6 each.

**MADAME RAVARY** (H.T.). Golden yellow, shaded orange. 1/6 each.

**MARGARET DICKSON HAMILL** (H.T.). Straw-coloured, edged and flushed with delicate carmine. 1/6 each.

**MILDRED GRANT** (H.T.). Ivory white, flushed with pale peach. 1/6 each.

**MISS MAY MARRIOTT** (H.T.). Sport from "Daily Mail," colour a beautiful terra-cotta, sweetly scented and a strong grower. 1/6 ea.

**MISS WILLMOTT** (H.T.). Colour soft sulphury cream with the faintest flush towards the edges. 1/6 each.

**MOLLY SHARMAN CRAWFORD** (T.). Beautiful cardinal white in the bud stage, and the flower develops to satiny white. 1/6 each.

**MRS. A. R. WADDELL** (H.T.). Rosy scarlet, bud opening reddish salmon. 1/6 each.

**MRS. A. TATE** (H.T.). Coppery red, long perfect pointed blooms. 1/6 each.

**MRS. CHARLES E. PEARSON** (H.T.). Orange, flushed red, apricot and fawn, a delightful combination of colour. 1/6 each.

**MRS. CHAS. E. RUSSELL** (H.T.). Rose-carmine, with centre of rose-scarlet, fragrant; a large flower of good form. 1/6 each.

**MRS. O. E. SHEA** (H.T.). Brilliant madder-red shot with a glowing scarlet. The outer petals show a deep rose shading resting on an orange base. 1/6 each.

**MRS. CHARLES LAMPLOUGH** (H.T.). Soft lemon chrome, sweetly scented, a very superb variety. 2/6 each.

**MRS. DAVID McKEE** (H.T.). Creamy yellow; large, full and free. 1/6 each.

**MRS. H. R. DARLINGTON** (H.T.). A fine, exhibition rose, creamy yellow colour, perfect shape. 2/6 each.

**MRS. J. LAING** (H.P.). Flowers large; a beautiful soft pink. 1/6 each.

**MRS. WEMYSS QUIN** (H.T.). Intense lemon-chrome, washed with delicate maddery orange, which, when the bloom fully opens, becomes deep canary yellow; perfume delicious. Absolutely distinct. 1/6 each.

**MRS. C. V. HAWORTH** (H.T.). A beautiful decorative rose, colour a lovely combination of pink, orange and yellow. 2/6 each.

**MRS. GEORGE MARRIOTT** (H.T.). A most distinct and charming rose. The flowers are very large and absolutely perfect in shape and form. The colour is a deep cream and pearl, pencilled and suffused rose and vermilion; sweetly scented. 2/6 each.

**MRS. HENRY BALFOUR.** A greatly improved Mme. de Watteville; free from mildew; colour flesh pink, edges of petals deep pink. Gold Medal N.R.S. 2/6 each.

**MRS. HENRY MORSE** (H.T.). The whole flower has a clear sheen of bright rose deeply impregnated and washed vermilion veining on petals; fine for exhibition, bedding and massing; sweetly scented. Gold Medal N.R.S. 2/- each.

**MRS. REDFORD** (H.T.). Colour bright apricot orange, far the most striking variety in this lovely and pleasing tone of colour; very free and sweetly scented. 2/6 each.

**MURIEL DICKSON** (H.P.). Deep reddish copper in bud, paling with age to cherry red. 1/6 each.

**OPHELIA** (H.T.). Salmon-flesh shaded with rose, large and of perfect shape. 1/6 each.

**PHARISAER** (H.T.). Rosy white shaded salmon, bud long, vigorous. 1/6 each.

**RAYON D'OR** (Pernetiana). Pure yellow resembling Persian Yellow; the buds are orange yellow with crimson flush, the open flower is clear yellow. 1/6 each.

**RED LETTER DAY.** Brilliant glowing scarlet-crimson, cactus-shaped flowers; 1/6 ea.

**REINHARDT BÆDECKER.** A large yellow rose produced from Druschki and Rayon d'Or. 2/- each.

**RICHMOND** (H.T.). Pure bright scarlet, similar to "Liberty," but larger, fuller, and a better grower. 1/6 each.

**SEVERINE** (Pernetiana). Beautiful coral red fading to prawn red when fully expanded, lovely at the opening bud. 2/- each.

**SIR ROWLAND HILL** (H.P.). Dark velvety purple, very fragrant. 1/6 each.

**SUNBURST** (H.T.). Long pointed bud; flower cupped form; yellow with orange-yellow centre. 1/6 each.

**SOUVENIR DE GEORGE BECKWITH** (Pernetiana). Shrimp pink bud tinted orange yellow, very large full globular flowers on stiff stems, colour shrimp pink, tinted chrome yellow with deeper yellow at base of petals. 2/6 each.

**THE QUEEN ALEXANDRA** (H.T.). A startlingly brilliant flower of intense vermilion colour, deeply shaded old gold on reverse of petals, which spring from a pure orange base. 2/- each.

**VICTORY** (H.T.). Glowing scarlet, blooms long and pointed and of good form. Awarded Gold Medal N.R.S. 2/6 each.

**W. C. GAUNT** (H.T.). Colour velvety vermilion, tipped scarlet, reverse of petals crimson maroon, a good bedding variety. 1/6 each.

**WHITE KILLARNEY** (H.T.). A pure white sport from the well-known old pink. 1/6 each.

## DWARF ROSES IN POTS.

We also have a fine assortment of Dwarf Roses in Pots.
Our selection only, 2/6, 3/6 and 4/6 each. Packing and carriage extra.

# STANDARD AND HALF-STANDARD ROSES.

Unless otherwise priced, Standards 5/6 each, 60 - doz.; Half-Standards 4 6 each, 48/- doz.

Please note we cannot supply Standards and Half-Standards in any varieties other than those listed.

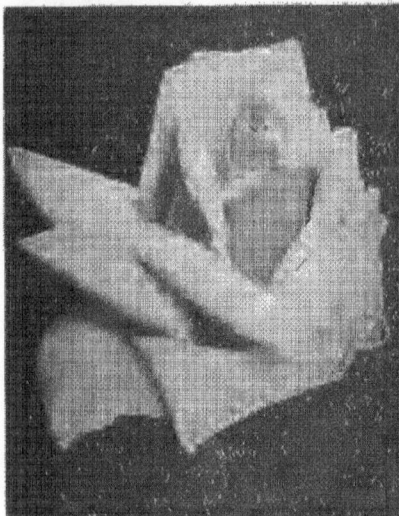

ROSE—LADY HILLINGDON.

*ADOLPH KARGER. Chrome yellow.
BETTY. Coppery rose.
BRITISH QUEEN. Pure white.
CAROLINE TESTOUT. Bright satiny rose.
CHATEAU DE CLOS VOUGEOT Red velvety scarlet.

CHEERFUL. Soft orange flame.
CHRISTINE. Deep yellow.
*COLONEL O. FITZGERALD. Blood red velvety crimson.
*COLUMBIA. True pink. 6 6 each.
*COUNTESS CLANWILLIAM. Delicate peach purple.
COVENT GARDEN. Rich deep crimson.
DAILY MAIL. Shrimp pink.
DUKE OF EDINBURGH. Bright vermilion.
*EMMA WRIGHT. Pure orange. 6 6 each.
FRAU KARL DRUSCHKI. White.
GENERAL McARTHUR. Dark velvety scarlet.
GEORGE DICKSON. Velvety black crimson.
GOLDEN EMBLEM. Beautiful yellow.
GOLDEN OPHELIA. Golden yellow.
GORGEOUS. Deep orange yellow.
HUGH DICKSON. Brilliant crimson.
IMPROVED CAROLINE TESTOUT. 6/6 ea.
LADY HILLINGDON. Golden yellow.
LADY PIRRIE. Deep coppery reddish salmon.
LADY ROBERTS. Rich apricot.
*LOS ANGELOS. Flame pink.
*LOUIS BALDWIN. Rich orange. 6/6 each.
*LYON ROSE. Salmon pink.
MADAME ABEL CHATENAY. Bright carmine rose.
*MARECHAL NIEL. Rich deep yellow.
MRS. J. LAING. Soft pink.
MRS. HENRY MORSE. Bright rose.
MRS. HERBERT STEVENS. White with peach shading.
*MRS. REDFORD. Bright apricot orange. 6 6 ea.
PHARISAER. Rosy white.
PRINCE CAMILLE DE ROHAN. Velvety crimson.
RED LETTER DAY. Brilliant scarlet -crimson.
RICHMOND. Pure bright scarlet.
SIR ROWLAND HILL. Dark velvety purple.
*SUNBURST. Yellow.
W. C. GAUNT. Velvety vermilion.

We have only a limited number of those marked with *.

# ROSES—TEA-SCENTED & NOISETTE.

The following prices are for bushes from the open ground.

LADY HILLINGDON (T). Fine golden yellow novelty, a cross between Papa Gontier and Lady Roberts, with long pointed buds, producing a glorious effect. 2/6 each.
MARECHAL NIEL (Noisette). Rich deep yellow, large, full and of perfect form; the petals are extra large, and of good substance, very sweet; a truly magnificent Rose. 2/6 ea.

GLOIRE DE DIJON. Yellow, shaded salmon, large, full, superb: one of the best. 2/6 ea.
MRS. HERBERT STEVENS (T). A hardy variety, faultless in shape, of exquisite grace and refinement, colour white. 2/6 each.
WILLIAM ALLEN RICHARDSON (Noisette). Fine deep orange yellow, very showy, and pretty for cutting. 2/6 each.

# CHINA ROSES AND AUSTRIAN BRIARS.

AUSTRIAN COPPER. Distinct and beautiful, golden terra-cotta colour, flowers single. 1/6 each.
FELLENBURG. Rosy crimson, very free bloomer. 1/6 each.

HERMOSA. Similar to Common China but more double, very effective. 1/6 each.
PERSIAN YELLOW. The deepest yellow, fairly full bloom, the most double of this class, 1/6 each.

# DWARF POLYANTHUS OR FAIRY ROSES.

These are of dwarf habit, producing compact clusters of charming flowers throughout the Summer and Autumn, they are ideal bedders, and most useful for edgings and borders, and they require very little pruning.

CECIL BRUNNER. Blush, shaded pale pink, small pointed buds.
EDITH CAVELL. Bright scarlet, white centre, a distinct and charming acquisition. Should not be confused with the H.T. variety.
ELLEN PAULSON. Dark pink, large, sweetly scented.
ERNA TERCHENDORF. Deep crimson, which does not fade ; very fine.
JESSIE. Bright orange red, in clusters.

MRS. W. CUTBUSH. Bright pink, distinct, most valuable bedding rose.
MADAME TURBAT. Flowers of a soft china rose colour.
ORLEANS. Clear rose colour, large trusses freely produced.
PHYLLIS. Bright cherry rose
RODHATTE. Clear cherry red
YVONNE RABIER. The best of all the white dwarf polyanthus ; an excellent grower.

1/3 each, 14/- per dozen, 110/- per 100.

# CLIMBING, PILLAR & WEEPING ROSES

**The following list of Climbing Roses includes the most beautiful and useful sorts in cultivation. Plants from the open ground 2/- & 2/6 each.**

STANDARD WEEPING ROSE, DOROTHY PERKINS.

**AMERICAN PILLAR.** Single rosy pink.
**AVIATEUR BLERIOT.** Saffron yellow flowers, in clusters, very fine. Vigorous.
**BLUE RAMBLER.** Steel blue, very distinct.
**BLUSH RAMBLER.** Beautiful soft blush.
**CLIMBING CAROLINE TESTOUT.** Bright satiny rose.
**CLIMBING IRISH FIREFLAME.** Deep madder orange splashed with crimson.
**CLIMBING K. A. VICTORIA.** Creamy white.
**CLIMBING LA FRANCE.** Bright silvery pink.
**CLIMBING LADY HILLINGDON.** Deep intense yellow. 3/6 each.
**CLIMBING MADAME ABEL CHATENAY** (new). Bright carmine rose, very fine. 3/6 ea.
**CLIMBING MME. A. CARRIERE.** White with a yellowish base; a rampant climber.
**CLIMBING MRS. W. J. GRANT.** Bright rosy pink.
**CLIMBING NIPHETOS.** Pure white.
**CLIMBING RICHMOND.** Pure bright scarlet very free-flowering.
**CRIMSON RAMBLER.** A splendid free-grower.
**DOROTHY PERKINS** (Hybrid Wichuriana). Clear soft pink flowers in large clusters.
**EMILY GRAY.** The finest golden Wichuriana Climbing Rose yet introduced.
**EXCELSA.** Red Dorothy Perkins.
**GLOIRE DE DIJON** (T). Buff, orange centre.
**GOLDFINCH.** Large clusters of yellow flowers.
**HIAWATHA.** Seedling from Crimson Rambler. Flowers are single, deep crimson, shading to white at base of petals.
**KEW RAMBLER.** Rose pink.
**LADY GAY** (Wichuriana). A brilliant and lovely shade of rose-pink.
**LADY GODIVA.** Soft pale flesh pink, like a carnation, a sport from Dorothy Perkins.
**MADAME A. CARRIERE.** Pure paper white.
**MARECHAL NIEL** (Noisette). Beautiful golden yellow of lovely form and delicious fragrance.
**MINNEHAHA** (Wichuriana). Deep rose, very double flowers produced in small panicles, very large trusses; extra fine.
**MONSIEUR DESIR.** Velvety crimson, shaded with violet, large and double.
**PAUL'S CARMINE PILLAR.** Bright rosy carmine, single flowers; charming.
**PAUL'S SCARLET CLIMBER.** Flowers vivid scarlet shaded with crimson; of strong climbing habit, with bright shiny foliage, flowering in great profusion, very large clusters, good sized semi-double blossoms, extra fine for Pillars and Pergolas.
**PEMBERTON'S WHITE.** Pure paper white.
**REINE MARIE HENRIETTE.** Deep lake red.
**WHITE DOROTHY PERKINS.** A white sport from the well-known Dorothy Perkins.
**WILLIAM ALLEN RICHARDSON** (Noisette). Fine deep orange yellow, very showy.
**ZEPHIRINE DROUHIN.** Bright carmine pink.

## STANDARD WEEPING ROSES.

The varieties named below can be supplied on tall Standards. These are the very best and most suitable varieties for this purpose. All who have seen these charming subjects will at once acclaim their worth.

| | |
|---|---|
| AMERICAN PILLAR. | GOLDFINCH. |
| ALBERIC BARBIER. | HIAWATHA. |
| *AVIATEUR BLERIOT. | *LADY GAY. |
| CRIMSON RAMBLER. | MINNEHAHA. |
| DOROTHY PERKINS. | PAUL'S SCARLET. |
| *EMILY GRAY | WHITE DOROTHY PERKINS. |
| EXCELSA. | |

4 to 5 feet stems, 7/- and 9/6 each.

We have only a limited number of those marked*

## CLIMBING ROSES (in Pots).

**We have a very fine lot of Climbing Roses in 6-inch pots, with shoots 6 to 10 ft. long, to bloom this season, of the following popular varieties. See description above.**

AVIATEUR BLERIOT. 4/- each.
CLIMBING NIPHETOS. 4/- & 6/- each.
CRIMSON RAMBLER. 4/- each.
DOROTHY PERKINS. 4/- each.
EMILY GRAY. 4/- each.
EXCELSA. 4/- each.
GLOIRE DE DIJON. 4/- & 6/- each.
GOLDFINCH. 4/- each.
HIAWATHA. 4/- each.
MARECHAL NIEL. 4/- & 6/- each.
MINNEHAHA. 4/- each.
PAUL'S SCARLET. 4/- each.
PEMBERTON'S WHITE. 4/- each.
WHITE DOROTHY PERKINS. 4/- each.
W. A. RICHARDSON. 4/- and 6/- each.

## MOSS ROSES, &c. 1/6 each.

**BLANCHE MOREAU.** Perpetual, pure white, in clusters, well mossed.

**CRIMSON GLOBE (HYBRID MOSS).** Buds nicely mossed, flowers deep crimson, large, full and globular, very vigorous.

**COMMON MOSS.** Rosy blush.
**CRESTED.** Rose, beautiful.

PROVENCE OR CABBAGE ROSES—

**OLD PROVENCE.** Rose colour, fragrant.
**WHITE PROVENCE.** White, beautiful.

# HARDY ORNAMENTAL TREES AND SHRUBS.

As we anticipated, last Autumn and Winter we experienced a great demand for all kinds of Ornamental Trees and Shrubs for refurnishing the gardens and pleasure grounds which had perforce been neglected during the war.

The coming Season will doubtless see a continuance of this work, and we have made ample preparations for supplying the needs of our customers in this important department of our business.

So much depends on the work of planting Trees and Shrubs which have been regularly transplanted, and here we are able to say that the whole of our huge stock of Trees has been transplanted during the past two years; we can therefore with confidence ask intending planters to place their orders in our hands.

During the past twenty years many most beautiful subjects have been added to the list of Flowering and Ornamental Trees, and in the following pages all the most charming subjects are included.

ACER.

**ACER (Maple). COLCHICUM RUBRUM.** Very soft red-coloured leaves. Height 8 ft. 6/6 each.

" **DRUMMONDI.** Beautifu silver variegated leaves. Height 8—10 ft. 6/6 each.

" **NEGUNDO CALIFORNICUM AUREUM.** This is a specially good variety, with large, broad golden foliage. Standards, with 3—4 ft. stem, 4/6 each.

" " **ALBA VARIEGATA.** Silver. Leaf beautifully variegated. Pyramids, 3—4 ft. 3/- each. Standards 4/6 each.

**ANDROMEDA FLORIBUNDA.** Dwarf compact growing shrubs, pure white bell-shaped flowers. Height 15 inches. 3/6 each.

**ARBUTUS UNEDO (Strawberry Tree).** Beautiful evergreen shrub, bearing an abundance of Strawberry-like fruit in Autumn. 3/6 each.

**ASH (*Fraxinus*). EXCELSIOR PENDULA (Weeping).** This is a splendid weeping tree. Standards 7/6, 10/6, and 15/- each.

" **GOLDEN LEAVED (Weeping).** 7/6 each.

" " **BARKED (Weeping).** A very telling variety. 7/6 and 10/6 each.

**AUCUBA JAPONICA VARIEGATA.** Height 2 ft. 2/6 each. Larger specimens 3/6 each.

**AZALEA MOLLIS.** These are excellent for forcing, or for outdoor planting. Mixed Seedlings. Nice bushy plants well set with buds. Height 15 inches 3/- each, 33/- doz.

**AZALEA PONTICA.** Orange yellow, sweetly scented, very free. Height 18 inches 3/- each, 33/- per doz.

**AZARA MICROPHYLLA.** A pretty, small-leaved evergreen shrub, covered with orange-red berries in Autumn. Height 2—3 ft. In pots, 2/6 each.

**BAMBOO (Hardy), AUREA.** Stems yellow, very straight, erect, growing close round the base. 3/6 and 4/6 each.

" **FORTUNEII VARIEGATA.** Beautiful bright green leaves, striated white. 3/- each.

" **JAPONICA (Bambusa Metake).** Dark green, sharply pointed leaves. 3/- each.

" **SIMONII.** Tall, straight, slender stems, runs very freely at the root. 5/- each.

" **NIGRA.** Stems glossy black after the first year. 5/- and 7/6 each.

**BAY, SWEET (Laurel Nobilis).** Height 1½—2 ft. 3/6 each, 2—3 ft. 5/- each.

**BEECH (*Fagus*) ATROPURPUREA.** Height 7—8 ft. 6/6 each.

" **FERN LEAVED.** 6/6 each.

**BERBERIS ARISTATA.** Yellow flowers in Spring, with bright red bark in winter. 2/6 ea.

" **DARWINII.** One of the most beautiful, producing a profusion of bright orange flowers, succeeded by handsome purple berries; recommended for the shrubbery, and for forming clumps in the pleasure ground. 2/- ea.

" **STENOPHYLLA.** One of the best; very graceful, small evergreen leaf, and covered with bright yellow flowers. Height 2—3 ft. 2/- each.

BERBERIS AQUIFOLIA. *See page 30.*

## Hardy Ornamental Trees and Shrubs—*(continued).*

BERBERIS THUNBERGII. Pretty early blooming species, white flowers; leaves in Autumn are tinted crimson. Height 2—2½ feet. 2/- each.

„ VULGARIS (Common Barberry). 2—2½ ft. 1/6 each.

„ WILSONII. Good hardy shrub of dwarf growth, the foliage is highly coloured in Autumn, and its long spurs and coral red berries makes it very attractive. 2/6 each.

BIRCH (Betula). SILVER-BARKED. Most beautiful. Height 9—12 feet 3/- and 4/- each.

BOX (Buscus). JAPONICA AUREA. A beautiful dwarf-growing variety with richly-coloured golden leaves; one of the best for Winter bedding or as an edging to large beds. 9—12 inches 1/6 each, 18—24 inches 3/- each.

„ HANDSWORTHII. This is the best of the green varieties of Box, close growing and of erect habit, foliage very broad, dense, and of the deepest green. Height 2—3 ft. 2/6 each, 3—4 ft. 3/6 each. Larger trees 5/- each.

BOX EDGING. One Nursery yard plants 3 yards. 1/6 per Nursery yard.

BROOM (Cytisus). ANDREANUS. This is perhaps the most distinct and beautiful Broom yet introduced, it is covered with maroon-crimson and yellow flowers. In pots, 3/- each.

„ COMMON YELLOW. In pots, 2/6 each.

„ NIGRICANS. Bright yellow, dwarf, compact habit, free flowering. In pots, 3/- each.

„ PRÆCOX. A beautiful free-flowering early variety. In pots, 3/- each.

„ PURPUREUS. In pots, 3/- each.

„ SPANISH. In pots, 2/6 each.

„ WHITE PORTUGAL. Very free-flowering, one of the best, most effective if planted in a mass. In pots, 2/- each.

BUDDLEA GLOBOSA. Orange globose flowers. 2/6 each.

„ VARIBILIS VEITCHII. In pots, 2/6 each.

„ MAGNIFICA. 2/6 each.

CALYCANTHUS (Allspice). Maroon-coloured flowers, very fragrant. Height 3 ft. 2/6 each

CHERRY (Cerasus). JAMES H. VEITCH. Said to be the finest variety yet introduced, flowers double rosy pink. Standards 5/- each.

„ RHEXII FLORE PLENA. Double white Cherry. Dwarf Bush 3/- each, Standards 5/- each.

„ WATERII, Double Rose. Beautiful rose. Standards 5/- each.

CHIONANTHUS VIRGINICA (Fringe Tree). Pure white flowers, very fragrant. Height 2 ft. 2/6 each.

CHOISYA TERNATA. Lovely white sweet-scented flowers. 2/6 and 3/6 each.

CORNUS, MASCULA ELEGANTISSIMA. A most beautiful tree. Dwarfs 4/- each.

„ SPATHII AUREA. Dwarfs 2/- each.

COTONEASTER FRIGIDA. This variety makes an excellent hedge, covering itself with berries. Height 3—4 ft. 1/6 each, Standards 4/- each.

„ HORIZONTALIS. Fan shaped, branches suitable for rookeries. Open ground. 2 ft. 2/6 each.

„ MICROPHYLLA. Fine for rookeries or walls; one of the best. Height 12—18 inches. 2/- each.

„ SIMONSII. Tall growing variety, produces quantities of bright scarlet berries. 3—8 ft. 1/6 each.

CYTISUS DALLIMOREII. 4/- each.

„ PENDULA PURPUREA. Rosy purple pea-shaped flowers. Dwarfs 4/- each.

DAPHNE CNEORUM (The Garland Flower). Very sweet evergreen, trailing habit. 3/- each.

DAPHNE MEZEREUM. Fragrant purple flowers early in Spring, whilst quite leafless. 3/- ea.

„ „ ALBUM. Similar to above, but pure white and deliciously fragrant. 3/- ea.

DEUTZIA CRENATA. Single white, flowering during June and July. 1/6 each.

„ „ FL. PL. Beautiful double rose-coloured flowers. 1/6 each.

„ GRACILIS. Pure white flowers, most useful for forcing. 1/6 each.

„ LEMOINEI. Beautiful white variety. 1/6 ea.

„ SCABRA. 1/6 each.

ELDER (Sambucus). GOLDEN LEAVED. Height 2—3 ft. 1/3 each.

„ SERRATIFOLIA FOLIUS AUREUS. Magnificent fern-like leaf, bright golden in colour. Height 2 ft. 2/6 each.

ELM (Ulmus). CAMPESTRIS, OR ENGLISH. These are beautiful grafted trees, and splendid for parks or avenues. Height 7—8 ft. 2/6 each, 27/6 per doz.

„ „ LOUIS VAN HOUTTE. Golden. Standards 5/- each.

„ CAMPESTRIS WEEPING. The true Umbrella Elm, an excellent weeping variety, specially adapted for planting as specimens on lawns, or in public parks and cemeteries. 12/6 each.

• „ DAMPIERI AUREA. This is a splendid Golden Elm in habit and growth and colour. Height 6—8 ft. 4/- each.

„ „ AUREA ROSSELSII. A beautiful golden variety. 4/- each.

„ „ WHEATLEYII. A good Elm of upright growth. Height 10 ft. 3/6 each, 40/- per doz., 12 ft. 5/- each.

ERICA ALPORTII. The best crimson. 1/6 each, 15/- per doz.

„ CARNEA. Abundance of reddish flesh-coloured flowers in March and April. 1/6 each, 15/- per doz.

„ „ ALBA. White variety of the preceding. 1/6 each, 15/- per doz.

ELDER.

## Hardy Ornamental Trees and Shrubs—*(continued)*.

HOLLY.

**ERICA HAMMONDII.** The lucky White Heath. 1/6 each, 15/- per doz.

**ESCALONIA MACRANTHA** Fine for seaside planting, neautiful foliage, covered with red flowers. In pots, 2/6 each.

„ „ **INGRAMI.** Of erect growth, pink flowers. In pots, 2/6 each.

„ **LANGLEYENSIS.** Long slender branches, producing numerous small branches, each bearing flowers of a bright rose carmine colour, with dark lustrous oval green leaves. 3/- each.

**EUONYMUS EUROPÆUS** (Spindle Tree). Very pretty in the Autumn when the fruit is ripe. Height 2—3 ft. 2/- each.

„ **JAPONICUS.** Green leaf variety. 1½—2 ft. 2/6 each.

„ „ Gold variegated. 18 inches 3/- each.

**FORSYTHIA SUSPENSA.** Suitable for wall or rockwork, flowers very early. 1/6 each.

„ **FORTUNEI.** 1/6 each.

„ **VIRIDISSIMA.** 2/- each.

**FUCHSIA RICCARTONI.** One of the hardiest and prettiest of outdoor kinds. 1/3 each, 12/6 per doz.

**GENISTA HISPANICA** (Spanish Gorse). Very free-flowering hardy shrub, flowers pale yellow. 2/- each.

**GORSE** (Ulex) **EUROPÆUS.** Double. In pots, 2/- each.

**GUM CISTUS** (Cistus Ladaniferus). A handsome shrub, growing to about 4 ft. In pots, 2/- each.

**HAZEL, PURPLE LEAVED.** Fine broad-leaved variety. 2/6 each.

**HOLLY, ARGENTEA MARGINATA.** Broad-leaved, silver, free grower and hardy. Height 2—3 ft. 7/6 each.

**HOLLY, GOLDEN QUEEN.** Ht. 2½—3 ft. 7/6 ea.

„ **REGINA (Silver Queen).** 2½—3 ft. 7/6 each.

„ **FRUCTO LUTEA.** Golden-fruited Holly Height 1½—2 ft. 7/6 each.

**HORSE CHESTNUT** (Æsculus). **SCARLET.** Splendid subject for avenues or for planting as park specimens. Height 9 ft. 3/- each. Specimen trees, 5/- each.

„ „ **DOUBLE WHITE.** Height 8 ft. 5/- each.

**HYDRANGEA PANICULATA GRANDIFLORA.** This is a beautiful plant and quite hardy, producing great drooping panicles of white flowers. 2—3 ft. 2/6 each.

**KERRIA JAPONICA.** Very pretty yellow shrub, free flowering. 1/6 each.

„ „ **FLORA PLENA.** A double variety of the above. 1/6 each.

„ „ **VARIEGATA.** 2/- each. 7/6 each.

**LABURNUM ADAMI** (Purple). Height 8 ft.

„ **VOSSII.** Distinct yellow flowers. Bush plants 2/6 each, Standards 3/6 and 5/- each.

**LAVENDER, COMMON (Lavendula).** 8d. each, 7/- per doz., 54/- per 100.

**LAUREL (Laurus).** **CAUCASICA.** Rich green foliage. Height 2—3 ft. 1/6 each, 15/- doz.

„ **COLCHICA.** This is one of the best varieties for the seaside. Height 2—3 ft. 1/6 each, 15/- per doz.

„ **LATIFOLIA.** Fine broad-leaved variety. Height 3—4 ft. 2/- each, 21/- per doz.

„ **ROTUNDIFOLIA.** Height 1½—2 ft. 1/6 each ; 15/- per doz. 2—3 ft. 21/- per doz.

„ **PORTUGAL (Lusitanica).** Height 3—4 ft. 3/6 each. Clipped Specimens 7/6 each.

**LAURESTINUS.** Height 2 ft. 2/6 each.

**LILAC** (Syringa) **ALBA.** The common single white Lilac. Height 3—4 ft. 2/6 each.

„ **CHARLES X.** Single deep purple Lilac, extra fine trusses. 2/6 each.

„ **LEMOINEI.** A variety with full double flowers of pale mauve colour. 3/- each.

„ **MADAME LEMOINE.** Double, compact, spike of the purest white. 4 ft. 3/- each.

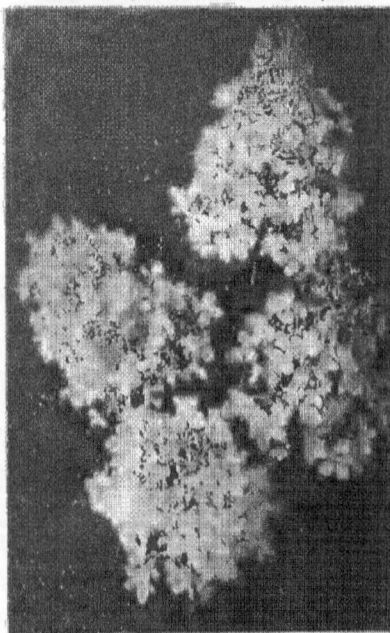

HYDRANGEA PANICULATA GRANDIFLORA.

## Hardy Ornamental Trees and Shrubs—*(continued).*

**LILAC, MARIE LEGRAYE.** Single white, very large flowers. 2—3 ft. 2/6 each, 3—4 ft. 3/6 each.

" **MICHAEL BUCHNER.** Rosy lilac, double, compact spikes. 3/6 each.

" **SOUVENIR DE LOUIS SPATH.** Dark red, large single flowers. 3/6 each.

" **VULGARIS.** The common single purple Lilac. Height 2—3 ft. 1/6 each, 3—4 ft. 2/6 each.

**LILACS.** Standards in variety. 4 ft. Stems 10/6 each.

**LIMES, Red Twigged, from Layers.** Standards 2/6, 3/6, 5/- and 7/6 each.

**LIQUIDAMBAR STYRACIFLUA (Sweet Gum).** Leaves very fragrant, and thrives well on damp situations, requires close pruning, flowers freely when established. Height 3—4 ft, 3/- each, 6—8 ft. 4/- each.

**MAPLES, JAPANESE** (*Acer*). **ATRO-PURPUREA.** In pots, 5/- and 7/6 each.

" **NORWAY PLATANOIDES.** This is a beautiful tree, very effective. 8—10 ft. 3/6 each.

" " **FOLIUS AUREUS.** A beautiful type of Golden Norway Maple. Dwarfs 3/- each, Standards 5/- each.

" " **SCHWEDLERI.** Leaves bronzy-purple, becoming as they mature a purplish green. Standards 5/- each.

**MAGNOLIA CONSPICUA.** White, reverse of petals suffused with purple. Height 2 ft. 5/6 each.

" **GRANDIFLORA (Evergreen).** Large white flowers and very fragrant, requires a south aspect. In pots, 2—3 ft. 6/6 and 7/6 each.

" **SOULANGEANA.** Very fine sweetly-scented variety, bearing purplish tinted flowers with white centre. Height 3 ft. 7/6 each.

**MOCK ORANGE (Philadelphus). AVALANCHE.** Very free flowering, sweetly scented. 1/6 each, 15/- per doz.

" **CORONARIUS.** Useful for shrubberies; flowers freely. 2—4 ft. 1/6 each, 15/- doz.

" **FLORA PLENA.** Large double white flowers. 2—4 ft. 1/6 each, 15/- per doz.

" **LEMOINEI.** Garden Hybrid. 2—4 ft. 1/6 each, 15/- per doz.

" " **BOULE D'ARGENT.** Dwarf habit, flowers very large. 2—4 ft. 2/- each, 21/- doz.

" " **GERBE DE NEIGE.** Very large white flowers and sweetly scented. 2—4 ft. 2/- each, 21/- per doz.

**MOUNTAIN ASH.** Standards 8—10 ft. 2/6, 3/6 and 5/- each.

**OAK, COCCINEA (Scarlet Oak).** The Scarlet Oak is one of the finest of our ornamental trees, the Autumn tints of the foliage being simply gorgeous. Height 8 ft. 3/6 each.

" **ILEX (Evergreen).** Grown specially in pots. Height 3—4 ft. 4/6 each, 4—5 ft. 6/6 each.

**OLEARIA HAASTII.** A dwarf evergreen smoke-resisting shrub, covering itself with masses of grey bloom, fragrant and lasting a considerable time. Perfectly hardy. Height 12—18 inches. 2/- each.

**OSMANTHUS ILLICIFOLIUS.** Height 1½—2 ft. 2/- each.

**PLANE, LONDON.** This is a splendid tree for street planting. Height 8—10 ft. 4/6 each. Specimens, extra good heads, 6/6 each.

**POPLAR TREMULA, WEEPING.** A very fine tree for lawns. 10/6 each

**PRIVET JAPONICUM.** Large, broad, shining green foliage. Height 2—3 ft. 2/6 each.

" **OVALIFOLIUM FOLIUS AUREUS (Golden-Leaved Privet).** This is the most showy hardy plant extant, foliage broadly margined with bright gold ; useful alike for hedges, or for planting singly in borders it cannot be surpassed. Height 1½—2 ft. 2/6 each, Selected Standards 6/6 each.

LILAC.

**PRUNUS PISSARDI (Purple Leaf Plum).** A handsome foliage tree with white flowers, quite hardy. Bushes 3—4 ft. 3/6 each. Standards 5 ft. stems and good heads, 4/6 each.

**PYRUS FLORIBUNDA.** Dwarfs 4—5 ft. 2/6 each. Standards 6—7 ft. 6/6 each.

" **ATROSANGUINEA.** Pyramids 5—6 ft. 3/6 each. Standards 5 - each.

" **MALUS. SIBERIAN Crab, Red.** 3/6 each. Standards 5/6 each.

" " **DARTMOUTH Crab.** Very fine. Standards 5/6 each.

" " **JOHN DOWNIE (Crab Apple).** A beautiful variety. Standards 6/6 each.

" **SORBUS LUTESCENS.** A noble tree with silvery leaves, pure white underneath ; bears large red berries when a large tree. Standards 5/- each.

**RHODODENDRONS.** Choice Grafted Named Hybrids. We are desirous of directing particular attention to this gorgeous class of plants. We strongly recommend them to our Customers, as the different gradations in colour gives a very pleasing effect. Well budded bushy plants, 6/- each, 66/- per doz. Choice Hybrid Seedlings. Height 15—18 inches 21/- per doz., 18—24 inches 27/6 doz.

" **PONTICUM.** Good bushy stuff, well rooted. Height 12—15 inches 15/- doz., 100/- per 100. It is generally believed that Rhododendrons cannot be grown without having a rich peaty soil, our experience however shows that suitable materials can be produced where good peat soil cannot be procured ; the top soil of old pasture land, mixed with leaf-mould and well rotted manure, forms an excellent and suitable compost

**RIBES AUREUM (Flowering Currant).** Flowers very early, bright golden yellow. Height 3 ft. 1/3 each.

" **SANGUINEUM.** Height 2—3 ft. 1/6 each.

" " **ALBIDUM.** White shaded with pink. Height 2—3 ft. 1/6 each.

" " **FLORA PLENA.** Flowers double red. Height 2—3 ft. 2/- each.

**ST. JOHN'S WORT (Hypericum). CALYCINUM** Transplanted 7/6 per doz., 55/- per 100.

## Hardy Ornamental Trees and Shrubs—*(continued)*.

GOLDEN PRIVET.

**SKIMMIA JAPONICA.** Female variety. 1 ft. 3/- each.
" **FRAGRANS.** White, very sweet-scented. 3/- each.
**SNOWDROP TREE** (Halesia Tetraptera). Height 3—4 ft. 3/- each.
**SNOWY MESPILUS** (*Amelanchier*). CANADENSIS. Standards 4/6 each.
**SPIRÆA OPULUS AUREA.** The golden-leaved variety. 1/6 each.
" **ARGUTA.** Very early flowering, pure white variety. 1/6 each.
" **ARIÆFOLIA.** Vigorous grower, large trusses of cream flowers in July. 1/6 each.
" **CALLOSA, RED.** Free flowering. 1/6 each.
" " **ALBA.** Pretty white flowers. 1/6 ea.
" **CONFUSA.** Slender habit, capital for forcing. 1/6 each.
" **DOUGLASSI.** Rose-coloured flowers, free-flowering variety. 1/6 each.
" **ANTHONY WATERER.** A splendid dwarf-growing variety, covered with deep crimson flowers and keeps in bloom a long time. 1/9 each, 18/- per doz.
" **AITCHISONII.** Long fern-like foliage, red bark, long panicles of pure white flowers. Extra good. 1/6 each.

**SPIRÆA LINDLEYANA (Himalaya).** Very hold spreading habit, large foliage and white flowers. Height 2 ft. 1/6 each.
" **MENZIESI.** Lovely pink flowers. 2/- each.
" **SALICIFOLIA ALBA.** White flowers in early summer. 2/- each.
**SUMACH, VENETIAN** (Rhus Cotinus). Lovely foliage, which deepens in Autumn. 3 ft. 2/- each.
" **TYPHINA** (Staghorn's Sumach). Fine bold foliage. Height 4—5 ft. 1/6 each.
**SYCAMORE (Acer). VARIEGATA.** These are very beautiful trees and quite hardy. 2/6 & 3/6 each. Specimens 5/- each.
" **LEOPOLDI.** Beautiful purple flesh coloured. 4/6 each.
**TAMARIX GALLICA** (Common Tamarisk). Very slender, graceful habit. Height 3 ft. 1/- each. 10/6 per doz., 60/- per 100.
" **TETRANDRA.** Height 2—3 ft. 1/- each. 10/6 per doz.
**THORNS** (Cratægus). Double Crimson, Paul's, Pyramids 3/6 each. Standards 4/6 each.
" **SINGLE SCARLET.** Pyramids 3/6 each, Standards 4/6 each.
" **DOUBLE WHITE.** Standards 4/6 each.
**VERONICA BUXIFOLIA.** Good strong hardy variety. 1/6 each, 15/- per doz.
" **TRAVERSII.** This is a most useful variety, stands well when planted near the sea, and makes a compact bush. 1/6 each, 15/- doz.
**VINCA MAJOR.** Large leaved variety with blue flowers. 9d. each, 7/6 per doz.
" " **ELEGANTISSIMA.** With beautiful variegated foliage. 1/3 each, 12/- per doz.
" **MINOR.** Small leaves and blue flowers. 9d. each, 7/6 per doz.
**VIBURNUM CARLESI.** White shaded pale pink; sweetly scented. Height 1½—2 ft. 3/6 each.
" **OPULUS** (Snowball Tree). Beautiful flowering shrub with large globular white flowers. 3—4 ft. 2/6 each.
" **PLICATUM.** One of the best, a particularly showy shrub, flowers half the size of the Guelder Rose, pure white, splendid for planting on trellis or wall, forces well. Height 2—3 ft. 3/6 each.
**WEIGELIA AMABILIS.** Bright pink, strong grower and flowers freely. 1/6 each, 15/- per doz.
" **ABEL CARRIERE.** Rosy carmine, very free. 1/6 each, 15/- per doz.
" **ALBA.** Pure white. 1/6 each, 15/- doz.
" **EVA RATHKE.** A very beautiful variety, flowers dark red and carried in great quantities. 1/6 each.
**WILLOW** (Salix). AMERICAN WEEPING. A fine weeping variety. 7/6 each.
" **BABYLONICA** (Weeping Willow). Standards 6—8 ft. 5/- each.
" **KILMARNOCK.** This is a splendid Weeping variety. 7/6 each.
**YUCCA FILAMENTOSA.** Most useful plant, flowers freely. 2/6 each.
" **RECURVIFOLIA.** Particularly hardy variety. 3/6 each.

---

# SHRUBS suitable for Bedding and for Winter Decoration Outdoors.

We beg to draw your attention to the undermentioned Evergreen Shrubs, which we have pleasure in recommending for the purpose of winter decoration, and for the formation of permanent beds. By their compact habits, together with their capacity for being kept exceedingly dwarf, by being pruned to almost any extent, and by the many pleasing shades of colour that may be selected from amongst them, they provide both in winter and summer a most distinctive feature.

## CLEMATIS.

**DUCHESS OF ALBANY.** Beautiful bright pink. July. 2/6 each.
**DUCHESS OF EDINBURGH.** The best of all the double whites, deliciously scented. June to July. 2/6 each.
**GRACE DARLING.** A delicate tree of bright rosy carmine. 2/6 each.
**HENRYI.** Beautiful large creamy white. July to Oct. 2/6 each.
**JACKMANII.** Intense violet purple. July to Oct. 2/6 each.
**JACKMANII ALBA.** A grand pure white variety. 2/6 each.
**JACKMANII, RED.** A fine new red variety of true Jackmanii type. July to Oct. 2/6 each.
**JACKMANII SUPERBA.** Similar to "Jackmanii," but the colour more intense. July to Oct. 2/6 each.
**LADY BETTY BALFOUR.** A fine, deep, velvety purple. 2/6 each.
**LA LORRAINE.** Soft clear rose tinted with blush. July to Oct. 2/6 each.
**MARCEL MOSER.** Mauve violet with red bar, very fine. 2/6 each.
**MISS CRAWSHAW.** Distinct solferino pink. 2/6
**MONTANA—RUBENS.** Flowers a soft bright rosy lilac with white. Quite hardy. 2/6 ea.
**MONTANA GRANDIFLORA.** Large, pure white starry flowers. May and June. 2/- each.
**MRS. CHOLMONDELEY.** Lavender. May to July. 2/6 each.
**MRS. HOPE.** Satiny mauve. July to Aug. 2/6 ea.
**NELLY MOSER.** Light mauve, with bright red bars. July to Oct. 2/6 each.
**PRESIDENT.** Purple, suffused with claret, good. July to Oct. 2/6 each.
**VILLE DE LYON.** Bright carmine, red. July to Oct. 2/6 each.

# HARDY CLIMBING & OTHER PLANTS.

**These are mostly grown in pots, and can be supplied and planted any time of the year with perfect safety.**

**ABELIA RUPESTRIS.** A charming little wall shrub, with clusters of small white flowers, remaining in bloom for a considerable period. 3/6 each.
**AKEBIA QUINNATA.** Purplish brown. 2/6 ea.
**AMPELOPSIS** (Creepers).
,, **LOWII.** Small Palmata leaves. 2/- each.
,, **VEITCHII.** Small-leaved. 2/- each.
,, **PURPUREA.** Dark-leaved. 2/- each.
,, **Hederacea.** Virginian Creeper. 2/- each.
,, **Sempervirens.** Evergreen. 2/- each.
**ARISTOLOCHIA SIPHO.** Deciduous. 2/6 ea.
**AZARA MICROPHYLLA.** Beautiful. 3/- each.
**BIGNONIA RADICANS** (Trumpet Flower). 3/- ea.
**CEANOTHUS, GLOIRE DE VERSAILLES.** Panicles of sky-blue flowers. 3/- each.
,, **Azureus.** Pale blue. 2/6 each.
,, **Divaricatus.** Very pale blue. 3/- each.
,, **VEITCHII.** Blue. In pots. 3/- each.
**CHIMONANTHUS FRAGRANS.** Sweet. 2/6 ea.
**CRATÆGUS LÆLANDII.** Red-berried. 3/- ea.
**ESCALLONIA MACRANTHA.** Crimson. 2/6 ea.
**HOP,** Common Green-leaved. 1/6 each.
,, **GOLDEN-LEAVED.** Very rapid climber. 2/- ea
**IVY (Hedera)—Cavendishii.** Silver. 2/6 each.
,, **ALGERIENSIS VARIEGATA** (new). Large-leaved silver. In pots. 3/6 each.
,, **RUBRA MARGINATA.** One of the best small-leaved variegated Ivies. In pots, 3/6 each.
,, **ALBA VARIEGATA** (Crippsii variety). Fine silver variegated foliage, small leaved. In pots, 3/- each.
,, **Clouded Gold.** Fine. 2/6 each.
,, **CHRYSOPHYLLA.** Sulphur yellow. 2/6 ea.
,, **DENTATA.** Dark green foliage. 2/6 each.
,, **Emerald Green.** Glossy leaves. 2/6 each.
**IVY, IRISH.** Fine, quick-growing variety. Strong plants from open ground, 3 ft., 12/- doz. In pots, 4—6 ft., 2/- each, 21/- doz., 10 ft. 5/- ea.

**IVY (Hedera)-Palmata.** Handsome var. 2/6 ea.
,, **Tricolor.** Very pretty. 2/6 each.
**IVIES, Variegated and Green.** Strong staked plants. 3—4 ft. 2/6 each, 4—5 ft. 3/6 each, 6—8 ft. 5/- each.
12 Ivies, distinct, in pots, for 25/-. 6 varieties.
**JASMINUM** (Jasmine)—
,, **Nudiflorum.** Yellow, blooms in December and January. 2/6 each.
,, **Officinale.** White. 2/6 each.
,, **REVOLUTUM.** Evergreen. 2/6 each.
**KERRIA JAPONICA FL. PL.** Yellow. 2/6 ea.
**LONICERA** (Honeysuckle)—
,, **Aurea reticulata.** Golden-veined. 2/6 each.
,, **EARLY DUTCH.** 2/6 each.
,, **Flexuosa.** Evergreen. 2/6 each.
,, **HALLII.** White, evergreen. 2/6 each.
,, **Late Red Dutch.** Well-known. 2/6 each.
,, **SCARLET TRUMPET.** Fine scarlet flowers. In pots, 2/6 each.
**PASSIFLORA CÆRULEA.** Blue. 2/6 each.
,, **CONSTANCE ELLIOTT.** White. 2/6 each.
**PERIPLOCA GRÆCA.** Rapid climber. 3/- ea.
**POLYGONUM BALDSCHUANICUM.** A splendid free-flowering climber. 3/- each.
**PYRUS JAPONICA.** Valuable early Spring flowering plant, rich scarlet. 3/- each.
,, **ALBA.** White form of the above. 2/6 each.
,, **ATRO-COCCINEA.** Double-flowered. 2/6 ea.
,, **KNAPHILL SCARLET.** Bright vermilion, the best of this class. 3/6 each.
,, **MAULEI.** Red flowers in Spring, golden-yellow fruit in Autumn. 3/- each.
**VITIS, Coignetiae.** The Crimson Glory Vine. Brilliant scarlet in Autumn. 3/6 each.
,, **Purpurea.** Claret-coloured foliage. 3/6 ea.
,, **THUNBERGII.** Handsome. 3/6 each.
**WISTARIA MULTIJUGA.** 5/- and 7/6 each.
,, **SINENSIS.** Lilac-mauve. 5/- and 7/6 each.
,, **SINENSIS ALBA.** 5/- and 7/6 each.

**Choice Hardy Climbers, our own selection, including Roses and Clematises, 27/6 per doz.**

# TRANSPLANTED FOREST TREES,
## HEDGING AND UNDERWOOD PLANTS FOR GAME COVERTS.

One of the most important lessons which this war has taught us is the absolute necessity of growing more timber in the United Kingdom, and the Authorities are already embarking on large schemes of afforestation, but much can also be done on small areas of land. Even where the land is unsuitable for ordinary cultivation many kinds of valuable timber trees will thrive and give a good return on the outlay necessary for trees and planting.

SPECIMEN ROW OF LIME TREES IN OUR NURSERIES.

To meet the demand for Forest Trees we have considerably increased our Stock in this direction, and it will be found to comprise all the best varieties for Forest and Estate planting ; further, we make it a practice to grow our trees thinly in wide rows which ensures their being thoroughly hardened when moved to exposed situations. We shall be pleased to give estimates, free of cost, for all kinds of planting, Forest or otherwise, and will carry out the entire work if desired.

Our prices, taking quality into consideration, are exceedingly moderate, and we are prepared to compete with any respectable firms in the trade. Extra Selected Plants, and smaller quantities than quoted, will be charged proportionally higher.

We shall be pleased to submit samples of any Forest Trees.

**ACACIA, Common.** Very ornamental tree, growing freely on poor, sandy soil. Specimen Trees, 3/6 each.

**ARBOR-VITÆ** (Thuya Occidentalis). Very compact growing tree, hardy, makes a splendid ornamental hedge. Height 2-3 ft. 2/- each, 21/- per doz.; 3-4 ft. 3/- each. Larger trees 3/6 and 5/- each.

### ASH.

**COMMON.** A valuable and well-known timber tree of considerable beauty. The wood is white grained, very tough, and elastic, is used by wheelwrights and also for the making of various implements, and is generally in good demand. The trees may also be used as a covert plant if cut down every few years. Height 2—2½ ft. 8/- per 100, 60/- per 1000    „   3—4 ft. 10/- per 100, 75/-  „

**MOUNTAIN.** Very ornamental tree, most valuable for exposed situations. Height 3—10 ft. 2/6 & 3/6 each.

**AUCUBA JAPONICA VARIEGATA.** Grows well under shade of trees. Height 1½—2 ft. 2/6 each, 27/6 per doz.

**BERBERIS AQUIFOLIA.** Excellent for game coverts and shrubberies. Height 1—1½ ft. Transplanted 12/- doz.

„ **DARWINII.** A densely branched evergreen with large yellow flowers. Height 2—3 ft. 2/- each, 21/- per doz.

**BERBERIS VULGARIS.** The common fruit-bearing Barbery. Height 2—2½ ft. 1 6 each, 15 - per doz.

**BEECH.** Sizes and prices on application.

**BIRCH, Common.** Succeeds in moist situations, wet, undrained or bog land. Timber fine-grained and valuable. Makes the best charcoal for gunpowder. Height 10—12 ft. 3/6 and 5/- each.

„ **SILVER.** One of the most beautiful of our forest trees. Succeeds in moist situations, wet, undrained, or bog land. Height 9—12 ft. 3/- and 4/- each.

**BOX, TREE.** This forms a handsome hedge for ornamental purposes. Height 2—3 ft. 3/- each, 33/- per doz.; Specimens 3/6 each.

**BRIAR, SWEET.** Height 1½—2 ft. 1/- each, 10/6 per doz.

**BROOM, Common.** Transplanted, 5/- per doz., 35/- per 100.

**CHESTNUT, HORSE.** Very handsome flowering tree. Splendid for avenues or parks, requires good land. Height 6—8 ft. 2/6 each. Specimen Trees 5/- each.

„ **SPANISH.** Highly ornamental and rapid grower when planted in good soil. Requires a sheltered situation. Specimens 5/- each.

## Forest Trees—*(continued).*

SPRUCE.

**CUPRESSUS LAWSONIANA.** Most popular of all Cupressus introduced by Messrs. Lawson, Edinburgh, in 1854; forms a fine evergreen hedge. Height 1½—2 ft. 2/6 each, 2—3 ft. 3/6 each, 5—6 ft. 5/- each.

**ELDER, Silver variegated.** 2—3 ft. 1/- each. 10/6 per doz.

**ELM, ENGLISH.** This is a tall and elegant tree of rapid and erect growth. Standards 2/6 ea.

### FIR.

**DOUGLAS (Pseudotsuga Douglasii).** This very handsome Fir grows to a height of 100 to 180 feet, and forms a very graceful tree. It prefers a sheltered position and thrives best when planted some distance from the sea coast.

Twice transplanted.
Height 1½—2 ft. 30/- per 100, 240/- per 1000
2—3 ft. 40/- per 100.

**SCOTCH (Pinus sylvestris).** A well-known and very valuable tree, thrives best in peaty or sandy soil, and is extensively planted on very poor land. The tree grows to a height of 50 to 100 feet

Transplanted.
Height 1—2 ft. 15/- per 100, 120/- per 1000.

**SPRUCE (Abies excelsa).** One of the best known of all the Abies and generally planted for all purposes. Very hardy, and will thrive on almost any soil. Splendid for game coverts, and coming quickly into use is of great value for many purposes. Grows to a height of 80 to 100 feet, and is useful as a nurse to Larch and other Forest Trees. Height 9—12 inches. 11/6 per 100, 90/- per 1000

**SITKA SPRUCE (Abies menziesii).** A very hardy species grows to height of 60 to 100 feet, but unless grown in a moist position it looses its needles early and presents a bare appearance. Height 9—12 inches, 15/- per 100, 120 - per 1000

**HAZEL.** Ht. 2—3 ft. 15/- 100, 3—4 ft. 20/- 100.

**HOLLY, Green.** Forms a beautiful hedge, if planted in double rows, leaving 12 inches between each row; also well adapted for mixing with thorn. Height 1½—2 ft. 15/- per doz., 2½—3½ ft. 2/6 each, 28/- per doz.

**HYPERICUM CALYCINUM.** An excellent covert plant for growing under the shade of large trees, beautiful when in flower, and not liked by Rabbits. 7/6 doz., 55/- 100.

**IVY IRISH.** Open ground. 12/6 per doz.

**LABURNUM, Common.** Very handsome early-flowering tree, covered with yellow blossom. Height 5—6 ft. 2/- each. Specimens 3/6 ea.

## LARCH.
### (Larix Europœa.)

Larch takes a very high place amongst trees by reason of the tender green of the young foliage. Naturally the Larch forms a tall clean trunk, and as the wood is firm and tough the trunks are in great request for scaffold poles and other purposes It is planted in considerable numbers in all parts of the country for economic purposes as well as for its beauty as a plantation tree. Larch will grow on almost any soil, but is usually best on a light or medium loam. It grows to a height of 80 to 100 feet.

Twice transplanted.
Height 1½—2 ft. 11/6 per 100, 90 - per 1000
2—2½ ft. 15/- per 100, 120/- per 1000

## LIMES.

In great demand for Avenues, Street and Park planting.

**Red Twigged, from Layers.** Very ornamental tree when grown singly. We have an extra fine lot of these. Standards 2/6, 3/6, 5/- and 7/6 each.

**MYROBELLA.** Very Scarce.
Height 1½—2 ft. 30/- per 100.
2—2½ ft. 35/- per 100.
We have a limited quantity suitable for gapping Extra strong at 40/- per 100.

LARCH

### Forest Trees—(continued).

**MAPLE, NORWAY.** Specimen Trees 3/6 each.
**OAK, ENGLISH.** Specimen Trees 3/6 & 5/- ea.
**POPLAR ABELE, Silver.** Very valuable quick-growing tree, useful for exposed situations, with silvery leaf ; grows freely on the coast. Standards 2/6 each.
„ **BALSAM.** Large fragrant foliage. Height 10 ft. 2/- each. Larger Trees 3/6 each.
„ **BLACK ITALIAN.** The most rapid growing of all our forest trees, grows fully in most soil; invaluable for shelter. Transplanted Spring 1922. 4—5 ft. 1/6 each, 16/- doz., 5—7 ft. 2/- each, 21/- per doz. Specimens 2/6 and 3/6 each.
„ **DELTOIDEA AUREA.** Is becoming a favourite, grows very freely, with rich golden foliage. Standards, Height 8—10 ft. 5/- each.
„ **LOMBARDY.** Very ornamental upright growing tree, often introduced in Landscape with effect, grows well in almost any soil, and most useful for close planting to act as a block. Height 4—5 ft. 1/6 each, 16/- doz., 5—6 ft. 2/- each, 21/- doz., 6—8 ft. 2/6 each, 24/- doz. Specimen Trees 3/6 each.

## PRIVET.

### OVAL-LEAVED.

Privet succeeds in shade but grows most strongly in an open situation, and this variety is practically evergreen where sheltered. Hedges of it may be clipped twice a year : in Winter and in June or July.
Very scarce, only a limited quantity to offer.

Height 1½—2 foot, 35/- per 100
„ 2—2½ feot, 40/-  „

**PINE, AUSTRIAN.** The Austrian Pine stands sea exposure well. Besides its great value for planting as an ornamental tree, it is one of the best for shelter. Succeeds on high, dry exposed situations, and on sea-shore. All well transplanted trees. Height 2—3 ft. 2/6 each, 27/6 por doz., 3—4 ft. 3/6 each.
**SEA BUCKTHORN.** 2½—3 ft. 12/- per doz.

SCOTCH FIR -(PINUS SYLVESTRIS).

ENGLISH YEW.

**SNOWBERRY.** Height 2—3 ft. 15/- per 100.
**SYCAMORE.** Very hardy, and stands the sea winds better than most other trees ; grows freely in any soil, very ornamental, and excellent timber. A limited quantity. Height 2—3 ft. 15/- per 100. Specimen Trees 2/6, 3/6 and 5/- each.
**THORNS or QUICKS.** Best for efficient hedges. Height 9—15 in., 12/- por 100, 100/- per 1000.
**YEW (Taxus), Common.** The most Ornamental of all evergreen hedges, it should not be planted in any situation accessible to animals that might eat it. Hardy and a compact grower. Height 1½—2 ft. 2/6 each, 27/- per doz., 2—2½ ft. 3/6 each, 40/- per doz., 3—3½ ft. 5/- each, 55/- per doz. Specimens 7/6 each.

## THE TRUE BAT WILLOW.

### (Salix alba *var.* cœrulea).

The fastest growing, most profitable and best variety.

This fine Willow is a first cross between *Salix fragilis* and *S. alba*, and is known as *Salix alba cœrulea*. It was at first thought to be only a variety of *S. alba*, but is found to grow much faster than that variety, and that it occurs only as a female plant.

It will thrive on moist, low-lying grass lands, by the side of ditches, or in fact on any good land with a fair depth of soil. It is of remarkably quick growth, and good three-year-old plants under fair conditions will in sixteen or eighteen years from the time of planting attain a marketable size for cricket but making, and should be worth from £5 to £10 per tree.

The true variety which we offer is grown from cuttings originally taken by Mr. Shaw, the eminent bat-maker, from a tree which he selected as being the best variety for making high-class cricket bats.

A limited quantity of Standards, 6—8 ft. 18/- per doz.

4—5 ft. Transplanted Spring 1922, 16/- doz.

# CONIFERÆ.

All have been recently transplanted so as to ensure as far as possible safety in removal, abundance of room and attention have been given so that each tree may be a perfect specimen, Our trees are all hand pruned with a knife, which is a most important point.

We offer well shaped plants, suitable for potting or window boxes, in good variety, our selection, 18 inches, 1/9 each, 21/- per doz.　24 inches, 2/3 each, 24/- per doz.

We have a fine lot of Dwarf-growing Coniferæ suitable for Rockeries at 3/6 each.

**ABIES CANADENSIS (Hemlock Spruce).** Very graceful tree, grows well and well worthy of a prominent sheltered place in the pleasure grounds. Height 2—3 ft. 3/- each.
„ **NOBILIS.** Very fast-growing Fir; hardy. Height 2—3 ft. 5/- each.
„ **NORDMANNIANA.** Height 3 ft. 5/- each.
„ **PINSAPO.** Height 2—3 ft. 5/- each.
**ARAUCARIA IMBRICATA.** A most distinct tree. Height 12—18 ins. 5/- each.
**CEDRUS ATLANTICA. Fast-growing.** Height 2—3 ft. 5/- each. 3—4 ft. 7/6 each.
„ **GLAUCA,** Glaucous-leaved. Height 2—3 ft. 5/-, 3—4 ft. 7/6 and 10/6 each.
„ **DEODARA.** 1½—2 ft. 5/-, 2—3 ft. 7/6 each.

CEDRUS DEODARA

**CRYPTOMERIA ELEGANS.** Plumose habit, very ornamental, with bronzy-crimson foliage in Winter. 2—3 ft. 5/- each.
**CUPRESSUS ALBA-SPICA.** Young, foliage tipped with white. Height 2 ft. 3/6 each.
„ **ALLUMII.** Very hardy, compact, upright habit, foliage rich silver blue. Height 1—2 ft. 3/6 each, 2—3 ft. 5/- each.
„ **ERECTA VIRIDIS.** Good Cupressus of erect habit. Ht. 1½—2 ft. 2/6, 2—3 ft. 3/6 each.
„ **LUTEA.** Height 1½—2 ft. 5/6 each.
„ **MACROCARPA (In pots).** Very handsome. Makes a very fine ornamental tree. 1/8 each, 16/- doz., £6 0s. 0d. per 100. From open ground, Transplanted. Height 2½—3½ ft. 3/6 each.
„ **NOOTKATENSIS.** Very pretty. Height 2—3 ft. 3/6 each.
**GINKGO (Maiden Hair Tree) (Salisburia adiantifolia).** Height 2—3 ft. 2/6 each.
**JUNIPERUS (Juniper). CHINENSIS.** Forms a beautiful bush of bright green foliage and very hardy. Ht. 2—2½ ft. 3/6 and 5/- each.
„ **COMMUNIS FASTIGIATA.** Irish Juniper. Height 2—3 ft. 2/6 each.
„ **SABINA.** Very hardy, well-known variety. Height 1½—2 ft. 2/6 each.
**PICEA PUNGENS.** Height 2—2½ ft. 2/6 each.

**PICEA PUNGENS GLAUCA (Blue Spruce).** Height 2 ft. 5/- each.
**PINUS CEMBRA.** Height 3—4 ft. 5/- each.
„ **EXCELSA.** Height 3—4 ft. 5/- each.
„ **STROBUS (Weymouth Pine).** A very hardy variety. 3—4 ft. 5/- each.
**RETINOSPORA OBTUSA.** Light green foliage. Height 3—4 ft. 3/6 each.
„ **PISIFERA.** Height 2—2½ ft. 3/6 each.
„ **AUREA.** One of the best golden shrubs. Height 3—4 ft. 5/- each.
„ **PLUMOSA.** Height 2—2½ ft. 2/6 each.
„ **AUREA.** Height 2 ft. 3/6 each.
„ **SQUARROSA.** Height 2—2½ ft. 3/- each.
**TAXUS AUREA (Golden English Yew).** Height 2 ft. 3/6 each, 3 ft. 5/- each.
„ **FASTIGIATA (The Irish Yew).** Very dark foliage, with upright habit. Height 2 ft. 3/6 each, 3 ft. 5/- each.
„ **AUREA (Golden Irish Yew).** Height 1½—2 ft. 2/6 each, 2—3 ft. 5/- each.
**THUJOPSIS DOLABRATA.** Compact habit. Height 1½—2 ft. 3/6 each.
**THUJA LOBBII.** Height 1½—2 ft. 2/- each, 2—3 ft. 2/6 each.
„ **LUTEA (Golden Arborvitæ).** Height 2—3 ft. 3/6 each.
„ **PLICATA AUREA.** A lovely golden shrub suitable for tubs or borders. 2—2½ ft. 3/6 ea.
„ **VERVAENEANA.** Rich golden shrub suitable for tubs or borders. Height 2½ ft. 3/6 each.

THUJA LOBBII.

# HARDY PERENNIAL FLOWERING PLANTS.

No part of the garden gives a greater return for the amount of time or money expended upon it than the Perennial Border. To ensure immediate effect we offer in many cases extra strong flowering clumps from the open ground.

ANCHUSA OPAL.

**ACANTHUS LONGIFOLIUS.** A robust plant, handsome foliage with long spikes of purple, rose, and white flowers. 1/- each.

**ACONITUM FISHERII.** Large heads of deep blue. 3½ ft. 1/6 each.

„ **NEPELLUS ALBUS.** White flowers. 3 ft. 1/- each.

**ACHILLÆA PTARMICA—THE PEARL.** Large double, pure white flowers ; fine for cutting. 1/- each.

**ADONIS VERNALIS.** Golden yellow with finely cut foliage. April-May. 1 ft. 1/- ea., 10/6 doz.

**AGAPANTHUS ALBA.** White. 2/6 each.

„ Blue. 2/6 each.

**AGROSTEMMA ATRO SANGUINEA.** Crimson with silvery foliage, 2 ft. 1/- each.

**ANCHUSA ITALICA,** Dropmore Variety. This is undoubtedly the finest blue-flowered herbaceous plant in existence. 1/3 each.

„ **ITALICA.** Sky blue flowers. 4 ft. 1/- each.

„ **MYOSOTIDIFLORA.** Small light blue flowers. 1 ft. 1/- each.

„ **OPAL.** Sky-blue variety, a pleasing shade of colour. 1/3 each.

„ **PICOTEE.** White edged with blue. 1/3 ea.

„ **PRIDE OF DOVER.** Best sky-blue. 3 ft. 1/6 each, 15/- per doz.

**ANEMONE, JAPONICA ALBA.** One of the best Autumn blooming plants. 1/- each.

„ „ **RUBRA.** Rosy red. 1/- each.

**ANEMONE PULSATILLA.** Pasque-flower. Rich purple stems and bracts covered with silky hairs. 1/6 each, 15/- per doz.

„ **QUEEN CHARLOTTE.** Large semi-double rose flowers. 1/3 each, 12/6 per doz.

„ **SYLVESTRIS PLENA.** Double snow-white flowers. 1/3 each.

**ANTHEMIS DANIELSII.** Yellow. 2½ ft. 1/- each, 10/6 per doz.

**ANTHEMIS PALLIDA.** Primrose yellow. 2½ ft. 1/- each. 10/6 per doz.

**ANTHERICUM GRAMINIFOLIUM.** Starry white flowers. 2 ft. 1/- each, 10/6 per doz.

**AQUILEGIA CHRYSANTHA.** Golden with outer petals flushed claret. 2½—3 ft. 1/- ea.

**AQUILEGIA, MRS. SCOTT ELLIOTT'S VARIETIES.** Mixed, long spurred. 1/- ea.

**ARTEMESIA LACTIFLORA.** White flowers. 4 ft. 1/- each.

**ASTERS (Michaelmas Daisies).** Beautiful Autumn bloomers.

„ **AMELLUS, BEAUTY OF RONSDORF.** Lilac pink. 2½ ft. 1/3 each, 12/6 per doz.

„ „ **KING GEORGE.** Flowers measuring 2 ins. across of deep violet. 2 ft. 1/3 each.

„ „ **FRAMFIELDI** Large violet blue flowers, very handsome. 2 ft. 1/- each. 2½ ft. 1/- each.

„ „ **PERRY'S FAVOURITE.** Soft pink.

„ **CLIMAX.** Clear light blue flowers. 1/- each.

„ **MULTIFLORUS.** Small white blooms on long slender sprays. 3½ ft. 1/- ea. 10/6 doz.

„ **NAMUR.** Soft pink. 4 ft. 1/- ea, 10/6 dz.

„ **NOVÆ ANGLIÆ.** Large bluish-purple flowers ; blooms in October. 1/- each.

„ „ **COL. FLETCHER.** Light rose-pink flowers. 5 ft. 1/- each.

„ „ **BELGEII DENSUS.** Blue. Height about 2 ft. 1/- each. 10/6 doz.

„ „ **ANTWERP.** Rose pink flowers. 5 ft. 2 6 each.

„ „ **BRUSSELS.** Pale lavender single flowers. 4 ft. 2/6 each.

„ „ **CREVIS.** White flowers. 2 ft. 1/6 each.

„ „ „ **ST. EGWIN.** Rose pink, compact habit, 3½ ft. 1/- each.

„ „ „ **GRACE MARY LEWIS.** Very large blue flowers. 3 ft. 1/- each.

„ „ „ **LIEUTENANT ROBINSON, V.C.'** (New). Bluish mauve double flowers. 3—4 ft. 2/6 each.

„ **DUMOSUS.** Mauve flowers. 1 ft. 1/- each.

„ **ERICOIDES, HON. EDITH GIBBS.** Graceful bunching sprays of small soft blue flowers. 3 ft. 1/- each.

„ **RUBELLA.** Late flowering. Bright rose. 4½ ft. 1/- each, 10/6 per doz.

„ **SAM BANHAM.** This is a white Climax, and is the best white yet introduced. 1/3 ea.

„ **SNOWDON.** A most charming variety, about 4 ft. high, bearing in Autumn lovely bracts of almost pure white flowers with pale yellowish disc. 1/- each.

„ **SUB CŒRULEUS.** Very large blue flowers on branching stems. Height 1½ ft. 1/- each.

„ **TOP SAWYER.** Clear blue, large flowers. Height 5 ft. 1/3 each, 12/6 doz.

„ **TRIUMPH.** Dark pink, late flowering. 5 ft. 1/3 each, 12/6 per doz.

„ **VIMICUS PERFECTUS.** White flowers. 3 ft. 1/- each.

We offer clumps of some varieties 2/- each.

**AURICULAS, ALPINE.** Choice seedlings, self-coloured and laced varieties. 9d. each, 8/- per doz.

**BUPTHALMUM SALICIFOLIUM.** Golden yellow flowers ; very showy. Height 3 ft. 1/- each.

**CAMPANULA GRANDIFLORA ALBA.** Very fine. 1/3 each, 12/- doz.

„ **LACTIFLORA.** Pale flowers of a light blue shade. 2-3 ft. 1/6 each, 15/- per doz.

„ **PERSICIFOLIA, Coronata Alba.** A most beautiful variety, growing about 3 ft. high. 1/6 each, 15/- doz.

„ „ **LINDLEYII** (new). Flowers of a pleasing light blue tinged inside with purple. 4 ft. Keeps in bloom from May to October. 3/6 each.

## Hardy Perennial Flowering Plants—(*continued*).

CAMPANULA PLATYCODON MARIESII.
Dark blue. 1 6 each.
„ PYRAMIDALIS ALBA. Tall white. 4 -5 ft.
1/- each.
„ PYRAMIDALIS (The Chimney Campanula).
Long spikes of blue salvia-like flowers, 1/-ea.
CATANANACHE CÆRULEA. Cornflower blue
flowers. 1/- each.
CENTAUREA DEALBATA. Rosy-purple
flowers, produced nearly the whole summer.
Fern-like foliage. 1/- each.
„ GLASTIFOLIA. Yellow flowers with
branching stems. 5 ft. 1/6 each.
„ MONTANA. Flowers blue. 2 ft. 1/- each
CHELONE BARBATUS. Bright red. 2½ ft.
1/- each.
CHRYSANTHEMUM MAXIMUM, STAR OF
ANTWERP. The best variety, producing
stiff stems and large flowers with broad
white petals. 3½—4 ft. 1/6 each, 15/- doz
„ „ KING EDWARD VII. One of the
finest varieties yet raised. It grows about
3 ft. high, and produces large, beautiful,
pure white flowers. 1/- each, 10/6 doz.
„ „ MRS. J. TERSTEIG. Flowers white.
2½ ft. 1/- each.
„ LEUCANTHEMUM. Flowers white. 3 ft.
1/- each.
„ SNOWFLAKE. A good white with broad
stiff petals. 2½ ft. 1/3 each.
CIMICIFUGA SIMPLEX. A handsome plant
with long erect spikes of feathery white
flowers. 5—6 ft. 1/- each, 10/6 per doz.
„ CORDIFOLIA. Flowers white. 3 ft. 1/- ea.
COREOPSIS GRANDIFLORA. Bright golden
yellow; blooms from June until September.
9d. each, 7/6 doz.
DELPHINIUM BELLADONNA. Lovely sky-
blue, one of the finest in cultivation.
2/- each.
„ „ SEMI-DUPLEX. Outer petals sky
blue, inner petals rosy mauve. 2/6 each.
„ KING OF DELPHS. Large flowers of
deep purple with small white eye. 2/6
„ LIZE VAN VEEN. This variety is remarkable by the immense size of the single
flowers, which have a pure blue colour.
2/6 each.
„ MOEREIMI. White. 2/6 each.
„ QUEEN OF LILACS. Outer petals pale
blue, inner petals deep lilac. 2/6 each.
„ QUEEN OF SPAIN. Semi-double, pale
blue and tinged lilac. 2/6 each.
„ REV. E. LASCELLES. Dark blue. 2/6
„ THE ALAKE. Semi-double, outer petals
dark blue, inner petals violet. 2/6 each.
OUR OWN SELECTION OF SORTS, 25/- doz.
„ SEEDLINGS from seed saved from best
named varieties, all of which have blossomed
this year. 1/6 each, 15/- per doz.
DICTAMNUS FRAXINELLA (Burning Bush).
This remarkable plant is one of the most
singularly interesting herbaceous perennials
in existence. 1/- each, 10/6 per doz.
„ ALBA. A fac-simile of the preceding,
but with pure white flowers; very showy.
1/- each, 10/6 per doz.
DORONICUM, HARPUR CREWE. A magnifi-
cent variety, bearing bold golden-yellow
flowers. 1/- each, 10/6 per doz.
DRACOCEPHALUM ALBUM. White flowers.
1/- each, 10/6 per doz.
ECHINACEA PURPUREA (Purple Cornflower).
Flowers crimson purple. 3 ft. 1/6 per doz.
EREMURUS HIMALAICUS. White on long
stems. 7 ft. 3/6 each.
ERIGERON SPECIOSUM SUPERBUM. Large
lavender blue flowers with yellow centre.
Height 3 ft. 1/- each. 10/6 per doz.
„ HYBRIDUS, B. LADHAMS. Deep pink
form of Speciosum. 1/3 each, 12/- per doz.
„ QUAKERESS. Large light blue. 1/- each.
ERYNGIUM ALPINUM (True). The finest of
the genus. The colour is a lovely metallic
blue. 1/- each, 10/6 per doz.

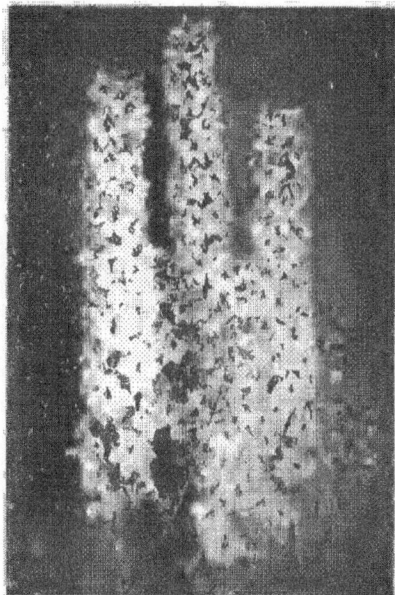

DELPHINIUM

ERYNGIUM GIGANTEUM. Ivory thistle, large
flat heads. 2 ft. 1/- each, 10/6 per doz.
„ PLANUM. Numerous small blue flowers,
useful for cutting. Height 3 ft. 1/- each.
FRANCOA APPENDICULATA. THE BRIDAL
WREATH. 1/- each.
FUNKIA LANCEOLATA. Deep green pointed
foliage. 9 ins. 1/3 each, 12/6 per doz.
„ SIEBOLDII. Flowers violet blue. 2 ft. 1/- ea.
GAILLARDIA. MRS. McKELLAR. Crimson
edged yellow. 1/6 each.
„ LE IDEAL. Yellow flowers, narrow ring of
crimson in centre. 1/6 each.
„ THE KING. Deep red, margin bright
yellow. 1/6 each.
„ CHOICE MIXED SEEDLINGS. 10/6 doz.
GALEGA OFFICINALIS. Numerous lilac
flowers on branching stems. Height 4 ft.
1/- each.
„ CARNEA FL. PL. Flowers double lilac.
4 ft. 1/- each, 10 6 per doz.
„ HARTLANDI. Flowers lilac. 4½ ft. 1/- ea.
GEUM, MRS. J. BRADSHAW. Double
scarlet. 9d. each, 8/- per doz.
GILLENIA TRIFOLIATA. Erect bushes
resembling a Spiræa in foliage, with white
flowers. 3 ft. 1/3 each, 12/6 per doz.
GYPSOPHILA PANICULATA. A fine border
plant, and most valuable for cutting.
1/- each, 10 6 per doz.
„ „ FL. PL. A pure white, double-flowered
of the well-known G. paniculata. 1/6 each.
HELENIUM PUMILUM MAGNIFICUM.
This forms immense heads of soft yellow
flowers, 2 to 3 inches across. 1/- each.
„ RIVERTON BEAUTY. Clear yellow; a
fine companion to Riverton Gem. 1/- each
„ RIVERTON GEM. Brilliant terra-cotta
red, a continuous bloomer from August to
end of October. 1/- each.
„ GRANDICEPHALUM. JULISONNE. Dark
yellow with dark brown centre. 3 ft. 1/-
each, 10/6 per doz.
„ BOLANDERI. Pale yellow. 1/- ea., 10/6 doz.

## Hardy Perennial Flowering Plants—(*continued*).

PÆONY.

**HELIANTHUS, MISS MELLISH.** Large single yellow. Height 5 ft. 1/- each, 10/6 per doz
,, **SOLEIL D'OR.** A fine variety, with deep orange-yellow double flowers. 1/- each.
,, **SPARSIFOLIUS.** Bright yellow. 1/- each.
,, **TOMENTOSUS.** The flowers are of a rich golden yellow colour and 3 inches across; very free bloomer. 9d. each.
**HELIOPSIS (Orange Sunflower). B. LADHAMS.** Flowers orange yellow. 4 ft. 1/- each.
,, **PATULA.** Flowers yellow. 4 ft. 1. - ea.
**HELLEBORUS NIGER.** Christmas rose. 2/- ea.
**HEMEROCALLIS, AURANTIACA MAJOR** Flowers large, trumpet shape, of a deep orange colour. Height 3 ft. 1/- each.
,, **FLAVA.** Large umbels of beautiful Lily-like flowers of a bright yellow colour. 1/- ea.
,, **FULVA.** Bronzy-orange. 1/- each.
,, **THUNBERGI.** Bright yellow. 1/- each.
**HESPERIS MATRONALIS FL. PL.** Flowers double white, fragrant. 1½ ft. 1/6 each.
**HEUCHERA BRIZOIDES.** A very distinct plant, forming tufts of dark bronzy foliage and splendid spikes of crimson flowers. 1/- ea.
,, **EDGE HALL HYBRID.** Large rose-coloured flowers. Height 18 in. 1/- ea.
,, **GRACILLIMA.** Flowers soft rose. 2 ft. 1/- ea.
,, **MICRANTHA.** White flowers.
,, **SANGUINEA SPLENDENS.** Crimson. 1/- ea.
,, **WOLLEY DODD.** ± Clear carmine flowers. 1 ft. 1/- each.
**IBERIS SEMPERVIRENS SUPERBA.** White. 1/- each.
**INCARVILLEA DELAVAYI.** Crimson. 1/- ea.
,, **GRANDIFLORA.** Rosy-crimson. 1/- each.
**INULA GLANDULOSA.** Orange. 1/- each.
,, **HELENIUM.** Yellow, large leaves and stiff stems, suitable for planting between shrubs. 4½ ft. 1/3 each.
**IRIS GERMANICA (Flag Iris.)** May be planted from September to March. Choice named varieties, our selection. 1/- each, 10/6 doz.
,, ,, **ADONIS.** Blue and purple. 1/- each, 10/6 per doz.

**IRIS GERMANICA ALBA.** Flowers large white, scented. 1/- each.
,, ,, **ALBICANS** syn. **PRINCESS OF WALES.** Flowers white. 1/- each.
,, ,, **ARC EN CIEL.** Yellow and white, slightly veined brown. 1/- each.
,, ,, **AUREA.** Yellow. 1/- each, 10/6 doz.
,, ,, **BLUE SKY.** Flowers mauve and blue. 1/- each.
,, ,, **DARIUS.** Flowers chrome yellow and purple. 1/- each.
,, ,, **FLAVESCENS.** Light primrose pencilled brown. 1/- each, 10/6 per doz.
,, ,, **GRACCHUS.** Yellow reticulated purple. 1/- each, 10/6 per doz.
,, ,, **MRS. H. DARWIN.** Snow-white veined purple. 1/- each, 10/6 per doz.
,, ,, **NEGLECTA.** Flowers pale blue and purple, reticulated white. 1/- each.
,, ,, **NE PLUS ULTRA.** Flowers chrome yellow and crimson. 1/- each.
,, ,, **NUDICAULIS.** Flowers violet. 1/- ea.
,, ,, **PALLIDA DALMATICA.** Lavender, large and fine. 1/- each, 10/6 doz.
,, ,, **PLURALIS.** Flowers light mauve and lavender blue. 1/- each.
,, ,, **QUEEN OF MAY.** Rosy-lilac. 1/- each, 10/6 per doz.
,, ,, **ROWLANDIANA.** Flowers white and veined purple. 1/- each.
,, ,, **VICTORINE.** Flowers white and blotched blue. 1/- each.
,, **FOETIDISSIMA** (The Gladwyn Iris). Flowers purple. 1/- each.
,, ,, **FOL. VAR.** Silver green variegated foliage. 1/- each.
,, ,, **KÆMPFERI.** Splendidly coloured. 1/- each, 10/6 per doz. Choice mixed. 1/- each, 10/6 per doz.
,, **PUMILA.** Purple. 18 inches. 1/- each, 10/6 per doz.
,, ,, **ALBA.** White. 18 inches. 1/- each, 10/6 per doz.
,, **OCHROLEUCA GIGANTEA.** Yellow. 3½ to 4 ft. 1/- each. 10/6 per doz.
,, **SIBERICA.** Pale blue, narrow foliage. 1/- each, 10/6 per doz.
,, ,, **SNOW QUEEN.** White. 1/- each. 10/6 per doz.
**LATHYRUS** (Everlasting Pea). **WHITE PEARL.** 1/- each.
,, **SPLENDENS.** Deep rose. 1/- each.
**LAVATERA OLBIA.** Rose pink. 1/3 each.
**LUPINUS ARBOREUS.** Yellow. 1/- each.
,, **ARBOREUS, SNOW QUEEN.** White. 1/- ea.
,, ,, **POLYPHYLLUS.** Blue. 1/- each.
,, ,, **ALBA.** White. 1/- each.
,, ,, **MOERHEIMI.** Rose. 1/- each.
,, ,, **ROSEUS.** Rose pink. 1/6 each.
**LYCHNIS CHALCEDONICA.** Scarlet. 1/- each.
**LYSIMACHIA CLETHROIDES.** A White Loosestrife. Fine large white flowers in gracefully drooping spikes, very handsome. 2—3 ft. 1/- each, 10/6 doz.
,, **VULGARIS** (yellow loose stripe). Flowers deep yellow. 2½ ft. 1/- each.
**LYTHRUM SALICARIA, PINK PERFECTION.** Rich pink flowers. 4 ft. 1/- each.
,, **VIRGATUM, ROSE QUEEN.** Bright rose flowers. 4 ft. 1/- each.
**MONTBRETIA, GEORGE DAVISON** (plants or clumps). Golden yellow. 1/- each.
**ŒNOTHERA ACAULIS VERA.** White. 1/- each.
**PAMPAS GRASS.** White plumes. 2/6 each.
,, **ROSE QUEEN.** Enormous plumes 7/6 each.
**PAPAVER, MRS. PERRY.** Orange-chrome. 1/- each.
,, **NUDICAULE.** Pale yellow, in pots. 9d. each.
,, ,, **MINIATUM.** Orange. 9d. ea., 8/- doz.
,, **ORIENTALE, ROYAL SCARLET.** The best scarlet. 1/- each, 10/6 per doz.
,, ,, **BOBS.** Salmon flowers. 3 ft. 1/- ea.
,, ,, **BRACTEATUM.** Large rich crimson flowers. 1/- each.
,, ,, **KING.** Flowers large crimson scarlet. 1/- each.

## Hardy Perennial Flowering Plants (*continued*).

PAPAVER ORIENTAL MAHONEY. Darkest crimson, well formed flowers. 1/- each.
" " MRS. J. HARKNESS. Salmon pink flowers. 1/- each.
" " MRS. MARSH. Flowers scarlet, striped white. 1/- each.
" " MRS. PEACOCK. Flowers rich salmon with dark blotch. 1/- each.
" " MRS. PERRY. Orange-chrome. 1/- ea.
" " PERRY'S WHITE. Flowers white with purple blotch. 1/- each.
" " PRINCE OF ORANGE. 1/- each.
" " PRINCESS VICTORIA LOUISE. Flowers charming pink. 1/-each.
" " SALMON QUEEN. 1/- each.
PÆONIES, HERBACEOUS. All the best varieties at 21/- per doz.
Choice sorts. 2/6 each, 27/6 doz.
PINKS. Choice named varieties. Our Selection. 1/- each, 10/- per doz.
PHLOX DECUSSATA (Late Phloxes)
" BARON VAN HEECKEREN (New). Salmon pink. 1/6 each.
" COQUELICOT. Brightest of scarlets, with purple eye. 1/- ea.
" DR. CHARCOT. Violet flowers, good truss. 1/- each.
" ELIZABETH CAMPBELL. Apple blossom pink. 1/3 each.
" EMILE KRANTZ. Rosy-lilac; a dwarf variety with dense heads. 1 ft. 1/6 each.
" ETNA. Brilliant scarlet, very fine. 1/- each.
" FRAU A. BUCHNER. The best white. 3 ft. 1/6 each.
" FRIEFRAULEIN-VON-LASSBERG. The purest and largest white Phlox. 1/6 each.
" G. A. STROHLEIN. Orange scarlet. 1/- ea.
" GOLIATH. Bright carmine with dark centre. 1/6 each.
" GRUPPEN KONINGEN. Pale rose. 1/- ea.
" IRIS. Violet blue, very fine. 1/- each.
" JULES SANDEAN (New). Clear ö pink flowers. 1/6 each.
" LEONARDE DE VINCI. Flowers white with vermillon centre. 1/- each.
" LE MAHDI. Violet blue. 1/6 each.
" M. HUGH LOW. Good crimson. 1/3 each, 12/- per doz.
" MISS E. WILLMOTT. Lilac, pale centre. 1/3 each, 12/- por doz.
" PANTHEON. Rosy salmon. 1/- each.
" RIJNSTROOM. Cerise, large heads. 1/- each, 10/6 per doz.
" SELMA. Pink with cherry red eye. 1/- each, 10/6 per doz.
" SINBAD. Flowers lilac mauve with carmine eye. 1/- each.
" TAPIS BLANC. The finest of all dwarf white Phloxes. 1 ft. 1/- each.
" TOREADOR. Flowers moss rose pink. 1/- ea.
PHYGELIUS CAPENSIS. Flowers bright red. 1/- each.
POLEMONIUM CÆRULEUM. 1/- each.
" RICHARDSONII ALBUM. Spikes of charming pure white flowers on stems of pale green foliage. 1 ft. 1/- each.
POTENTILLA, MIXED COLOURS. 1/- each, 10/6 per doz.
POTERIUM OBTUSUM. Rose pink, free flowering. 2 ft. 1/6 each.

PYRETHRUMS, BEST DOUBLE MIXED. 1/- each, 10/6 per doz.
" " SINGLE MIXED. 1/- each, 10/6 doz.
" SINGLE, AGNES MARY KELWAY. Rose, telling variety. 1/- each, 10/6 doz.
" " JAMES KELWAY. Scarlet. 1/- each, 10/6 per doz.
" DOUBLE, LA VESTULE. White tinged lilac. 1/- each.
" APHRODITE. Large white. 1/- each.
" MME. MUNIER. Pale pink. 1/- each.
" QUEEN MARY. Pink. 1/- each.
" To name. 10/6 per doz.
RANUNCULUS ACRIS FL. PL. Flowers double yellow. 2 ft. 1/- each.
RUDBECKIA CALIFORNICA. Yellow flowers on long stems. 5 ft. 1/- each.
" LACINIATA FL. PL. Double golden yellow. 1/- each, 10/6 doz.
" LÆVIGATA. Rich golden yellow flowers. 6 ft. 1/- each.
" NEWMANII. Yellow, black centre. 1/- ea.
SCABIOSA CAUCASICA. Lilac-blue. 1/- each.
" ELATA. Sulphur yellow. 1/- each, 10/6 doz.
SIDALCEA CANDIDA. Satiny white flowers. 3 ft. 1/- each.
" ROSY GEM. Bright rosy flowers. 1/- each, 10/6 doz.
SOLIDAGO, ALTISSIMA. Yellow. 1/- each.
" GOLDEN WINGS. Best golden yellow. 1/- ea.
" SHORTIA. Golden yellow. 1/- each.
SPIRÆAS, ARUNCUS. Plumes creamy white. 1/- each.
" FILIPENDULA FL. PL. Double white. 1/- ea.
" GIGANTEA. Feathery white flowers. 6ft. 1/6 each.
" PALMATA ELEGANS. Pale pink. Japanese Meadow Sweet. 1/3 each. 12/6 per doz.
" RIVULARIS. [Creamy white flowers. 6 ft. 1/6 each.
" THUNBERGII. White flowers. 3 ft. 1 6 ea.
" ULMARIA FL. PL. Double white. 3 ft. 1/6 ea.
STATICE LATIFOLIA. Sea Lavender. 18 inches. 1/3 each, 12/6 per doz.
THALICTRUM ADIANTIFOLIUM (*The Maidenhair Thalictrum*). Quite hardy. 1/- ea.
" AGUILEGIFOLIUM. Creamy white. 3 ft. 1/-ea
" DEPTEROCARPUM. Blue flowers. 2/- each.
" GLAUCUM. Yellow. 1/3 each, 12/6 per doz.
TRADESCANTIA VIRGINICA, "Flower-of-a-day." Flowers blue. 1¼ ft. 1/6 each.
" ALBA. White form of the above. 1¼ ft. 1/-
TRITOMA GOLDEN SPIRE. Fine foliage and long stems. 2½—3 ft. 1/6 each.
" NORTHIÆ. Scarlet & orange. 1/3 ea. 12/6 dz.
" TYSONII. Scarlet and orange, with bold glaucous foliage. 5 ft. 1/6 each.
" UVARIA GRANDIFLORA. Orange red. 1/- ea.
TROLLIUS ASIATICUS. Bright orange. 1/- ea.
" EUROPÆUS SUPERBUS. 1/- each.
" HIS MAJESTY. Round flowers of deep orange. 1/- each, 10/6 per doz.
" ORANGE GLOBE. Deep orange. 1/3 each.
TROPÆOLUM SPECIOSUM. Scarlet. 1/6 ea.
VERBASCUM CALEDONIA. 1/6 each.
" DENSIFLORUM. Golden bronze. 1/6 each.
VERONICA LONGIFOLIA SUBSESSILIS. Rich violet blue flowers. 3 ft. 1/- each.
VIOLAS. Choice named varieties. Our selection. 4d. each, 3/- doz., 21/- per 100.

We are able to supply extra sized clumps from the open ground, and we shall be glad to send a list of these to anyone wanting to obtain immediate effect.

# DANIELS' SPECIAL COLLECTIONS OF HARDY FLOWERING PLANTS

We have much pleasure in recommending the following four collections, which contain a very choice selection of the above, specially arranged for a brilliant and varied display of colour and a long continuance of bloom in the open garden.

COLLECTION A. 100 in 50 fine varieties, our selection, 65/-.

COLLECTION B. 50 in 50 fine varieties, our selection, 35/-.

Our own selection, 10/6 per doz.

COLLECTION C. 50 in 25 fine varieties, our selection, 30/-.

COLLECTION D. 25 in 25 fine varieties, our selection, 17/6.

Mostly clumps from open ground.

# CARNATIONS—Perpetual or Winter Flowering.

A beautiful free-flowering class for Winter and early Spring blooming under glass. Invaluable as cut flowers for Bouquets, Button-holes, &c. The plants we offer are all growing in 6½-inch pots and ready for blooming.

**AVIATOR.** Deep brilliant scarlet, free flowering variety.

**BARONESS DE BRIENEN.** Salmon pink, large flowers.

**BEACON.** Bright orange scarlet, large and free ; one of the best commercial scarlets.

**BELLE WASHBURN.** Scarlet.

**BRITISH TRIUMPH.** Rich bright crimson, well formed flowers of medium size and beautifully fimbriated, with excellent stem and calyx. The habit of growth is robust and healthy, flowering well into mid-winter.

**CAROLA.** The largest crimson ever raised, flowers of five inches diameter are no exception. A remarkably good keeper. Awarded many medals and special prizes all over Europe. Very strong grower. Should be cultivated entirely under glass.

**ENCHANTRESS SUPREME.** Clear pale salmon pink sport of " Enchantress ;" an improvement on this variety.

**LADY ALINGTON.** A delightful shade of rich rose salmon. The flowers are large and full, strongly scented.

**LADY FULLER.** Flowers large, deep salmon, plants healthy and vigorous.

**QUEEN ALEXANDRA.** Rich salmon-pink sport of the well-known " Scarlet Glow."

**ROSE PINK ENCHANTRESS.** Rose-pink sport of " Enchantress," good habit and a strong grower.

**SCARLET CAROLA.** Scarlet, very fine.

**TRIUMPH.** The best crimson market variety. The flowers are of the brightest possible crimson, and of large size. The shape leaves nothing to be desired, and the calyx never bursts. The habit of the plant is vigorous and healthy, and it produces flowers in large quantities.

**WHITE ENCHANTRESS.** White sport of " Enchantress." No better white can be obtained. Very good in every way.

**WIVELSFIELD WHITE.** Pure white, very fragrant.

Strong Plants, in 5 in. pots 2/6 each, 27/6 per doz. ; in 6 in. pots 3/6 each, 37/6 per doz.

# BORDER CARNATIONS.

Autumn planting is highly recommended.

This List is only a short one, but it contains all the very best varieties that are hardy and worthy of being grown.

**ARMISTICE.** Bright cherry red, non-splitting, on strong stout stems.

**CRIMSON QUEEN.** Deep crimson, strong grower, finely scented.

**GORDON DOUGLAS.** Deep crimson, sound calyx.

**MISS R. JOSEPHS.** Antique rose, strong grower.

**OLD CLOVE.** The finest scented deep crimson variety in existence

**PURITY.** A good white.

**RABY CASTLE.** Pink fringed edge.

**ROSY MORN.** Deep rose pink.

Strong Plants, 1/6 each ; 15/- per doz.

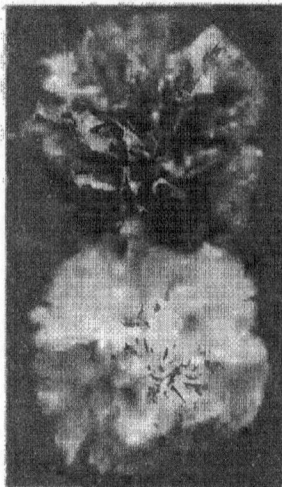

CARNATION.

# GARDEN PINKS.

**HER MAJESTY.** Large white ; very fragrant.

**MRS. SINKINS.** Large pure white flowers ; very fragrant.

Strong Plants, 1/- each ; 10/6 per doz.

# ALLWOOD'S HARDY PERPETUAL PINKS.

**DOROTHY.** Deep rose-pink with a rich dark centre.

**HAROLD.** Large double white.

**JEAN.** White with deep violet centre.

**MARY.** Pale rose pink, with light maroon centre.

**PHYLLIS.** Lilac, very free flowering and delightfully perfumed.

**ROBERT.** A delicate shade of old rose, with a light maroon centre.

In small pots 1/9 each, 18/- per doz.

# LILY OF THE VALLEY (Convallaria majalis).

For early forcing, single crowns of these should be planted about twelve in a five-inch pot, with the buds well above the surface. Cover the crowns with a little moss or an inverted flower-pot and place them in a good heat of say 85 to 90 degrees ; water frequently with tepid water, and if judiciously looked after they will bloom in four or five weeks from time of potting. Good single crowns are much the best for this purpose.

Selected Single Crowns for forcing    ..    ..     2/6 per doz., 17/6 per 100

# PLANTS SUITABLE FOR ROCKWORK.

We offer a splendid selection of all the most charming subjects for furnishing Rock gardens. We shall be pleased to supply detailed lists with advice as to planting if desired. Most of our stock is in pots and may be despatched at any season of the year.

Our own Selection, 7/6 and 9/- per doz.     55/- and 65/- per 100.

All the above are grown in pots.

# GREENHOUSE AND STOVE PLANTS.

ARAUCARIA EXCELSA.

**ARAUCARIA EXCELSA.** A fine plant for the conservatory, 3/6 and 5/- each.
**ASPARAGUS SPRENGERI.** 3/6 and 5/- each.
**ASPIDISTRA LURIDA.** 3/6 and 5/- each.
„ „ **VARIEGATA.** 5/- and 6/6 each.
**AZALEA INDICA.** We have a very choice collection, all good healthy flowering plants, varying from ten to sixteen inches across. 3/6 and 5/- each.
**BEGONIA, GLOIRE DE LORRAINE.** A profusion of bright pink flowers. 2/6 & 3/6 each.
**BORONIA MEGASTIGMA.** Lovely sweet-scented flower. 4/- each.
**CANNAS.** Dwarf varieties. 1/6 each; 15/- doz.
**CARNATIONS, MALMAISON VARIETIES.** In 5-in. pots, 2/6 each. Colours :—scarlet, yellow, and various shades of pink.
**CROTONS.** A fine collection of choice sorts in nice young plants. 5/- & 7/6 each.
**CYCLAMEN PERSICUM GIGANTEUM.** Strong Transplanted Seedlings. In 5-in. pots, 2/6 each, 27/6 per doz.
„ **PURE WHITE.** Very beautiful. In 5-in. pots, 3/6 each.
**DAPHNE INDICA ALBA.** Pure white, deliciously scented variety. 7/6 each.
**DAPHNE RUBRA.** Very sweet. 7/6 each.

**DRACÆNA AUSTRALIS.** Fine for furnishing. 2/6, 3/6 and 5/- each.
„ **GRACILIS.** Very useful for decorative purposes. 5/- each.
**ERICA HYEMALIS.** White. Lovely table plant. 3/6 and 5/- each.
„ **GRACILIS.** Pink. 3/6 and 5/- each.
**EUCALYPTUS CITRIODORA.** Deliciously scented. 2/- each.
**EULALIA GRACILLIMA.** Most elegant. 2/- each.
„ **ZEBRINA.** Very handsome. 2/- each.
**FERNS, GREENHOUSE.** A fine selection of the most useful and ornamental. 2/6 and 3/6 each.
**FICUS ELASTICA (India-rubber Plant).** 5/- ea.
„ „ **VARIEGATA.** Beautifully variegated with yellow. 7/6 each.
**GENISTA FRAGRANS.** Sweet-scented yellow flowers. 2/6 each.
**GLOXINIAS.** In beautiful variety. Strong plants. 21/- per doz.
„ **SEEDLINGS.** Very choice strain, ready in May. 10/6 per doz.
**GREVILLEA ROBUSTA.** A very ornamental plant. 2/- each.
**HOYA CARNOSA.** A charming stove climbing plant, producing wax-like flowers. 3/6 each.
**LAPAGERIA ALBA.** Lovely pure white, wax-like flowers ; beautiful. 5/- and 7/6 each.
„ **ROSEA SUPERBA.** Beautiful climber for the cool greenhouse. 5/- and 7/6 each.
**MYRTLES (Myrtus).** Nice young plants. 2/- ea. 21/- doz.
**OLEANDER (Nerium).** Pink and White. 3/6 each.
**PALMS.** A nice assortment of choice plants suitable for the dinner-table and general decorative purposes, including : — Areca sapida, Cocos Weddelliana, Kentia Belmoreana, Kentia Fosteriana. 5/-, 7/6, 10/6, 21/-, 31/- and 42/- each.
**PLUMBAGO CAPENSIS.** Blue. 2/6 each.
„ „ **ALBA.** 2/6 each.
**SOLANUM CAPSICASTRUM.** The well-known Solanum, with beautiful bright scarlet berries in Autumn and Winter. 2/6 each.
„ **JASMINOIDES.** A very pretty greenhouse climber. White flowers. 2/6 each.
**STEPHANOTIS FLORIBUNDA.** 5/- and 7/6 each.
**STREPTOCARPUS.** White, mauve and pink. These are a lovely strain. 2/6 each.
**SWAINSONIA GALEGIFOLIA ALBA.** White Pea-like flowers. 2/6 each.
**TACSONIA EXONIENSIS.** A fine variety. 3/6 each.
**GREENHOUSE PLANTS,** in choice variety, our selection, 35/- and 42/- per doz.

## SEEDLING GREENHOUSE PLANTS.

For those requiring an early display in the greenhouse, we can supply during August and September, strong, May-sown plants of the following, post free at prices quoted.

**ANTIRRHINUM, NELROSE.** A charming subject for winter decoration in the cool greenhouse. The colour is a lovely shade of silvery pink, with long spikes. Fine plants from cuttings. 3/6 per doz.
**CALCEOLARIA.** Choicest Mixed. Beautiful spotted and marked flowers. 3/- per doz.
**CINERARIAS.** Hybrida, large-flowered. 3/- per doz.
**CINERARIA STELLATA (Star Cineraria).** Very choice, tall growing. 3/- per doz.
**PRIMULA.** Crimson King. Splendid variety. 3/- per doz.

**PRIMULA.** Superb Blue. Very fine. 3/- per doz.
„ **Giant Pink.** Bright rosy pink. 3/- per doz.
„ **White Perfection.** The finest purest white. 3/- per doz.
„ **Daniels' Choicest Mixed.** Grand variety. 3/- per doz.
**PRIMULA MALACOIDES.** Pale Lilac flower. 3/- per doz.
„ **STELLATA (Star Primula).** Beautiful varieties, mixed. 3/- per doz.

In October the above can be supplied in 3-inch pots at 6/- per doz.
**Carriage extra.**

# PLANTS FOR SPRING BEDDING.

From the extended list of choice bulbs and plants we offer, selections may be made at a moderate cost that will provide quite a brave show of flowers from March to the middle of June. Wallflowers (especially the lighter coloured varieties) associate and contrast splendidly with the brilliantly coloured and taller growing sorts of the Darwin and May-flowering or "Cottage" Tulips (pages 50 & 53), and with the addition of a few clumps of Narcissi (pages 54 to 57), form delightful beds of colour, whilst rockeries and borders may be planted with such dwarf-growing subjects as Aubretias, Myosotis, Primroses, Polyanthuses, Daisies, Violas, etc., with highly satisfactory results.

## TRANSPLANTED SEEDLINGS.

ALYSSUM SAXATILE COMPACTUM.  per doz. 2·6

| | | | per doz. | per 100 |
|---|---|---|---|---|
| | | | s. d. | s. d. |
| BROMPTON STOCKS— | | | | |
| Purple | Rose | | | |
| Red | Scarlet | White | 1 6 | 10 6 |
| CANTERBURY BELLS— | | | | |
| Single Blue | .. | .. | .. 1 6 | 10 6 |
| Single Mauve | .. | .. | .. 1 6 | 10 6 |
| Single White | .. | .. | .. 1 6 | 10 6 |
| Calycanthema Blue | .. | .. | .. 1 6 | 10 6 |
| Double Blue | .. | .. | .. 1 6 | 10 6 |
| Double Rose | .. | .. | .. 1 6 | 10 6 |
| Double White | .. | .. | .. 1 6 | 10 6 |
| Double Mixed | .. | .. | .. 1 6 | 10 6 |
| Single Mixed | .. | .. | .. 1 6 | 10 6 |
| Calycanthema, mixed | .. | .. | .. 1 6 | 10 6 |
| „ | Rose | .. | .. 1 6 | 10 6 |
| „ | White | .. | .. 1 6 | 10 6 |

PANSIES.

| | per doz. | per 100 |
|---|---|---|
| | s. d. | s. d. |
| DIGITALIS (Foxglove)— | | |
| Spotted Mixed | .. 2 6 | 16 0 |
| Monstrosa | .. 2 6 | 16 0 |
| MYOSOTIS (Forget-me-nots)— | | |
| Indigo Blue | .. 2 0 | 14 0 |
| Sky Blue .. | .. 2 0 | 14 0 |
| Ruth Fischer. The best grown | 2 6 | — |
| PANSIES— | | |
| Giant Yellow | .. 2 0 | 14 0 |
| Prize Blotched | .. 2 0 | 14 0 |
| Giant Exhibition | .. 2 0 | 14 0 |

| | per doz. | per 100 |
|---|---|---|
| | s. d. | s. d. |
| PANSIES— | | |
| Madame Perret | .. 2 0 | 14 0 |
| POLYANTHUS— | | |
| Giant White | .. 2 6 | 16 0 |
| Mixed | .. 2 6 | 16 0 |
| PRIMROSES— | each. | |
| G. F. Wilson's Blue | 9d. | 7 6 — |
| SWEET WILLIAMS— | | |
| Giant White | .. 2 6 | 14 0 |
| Scarlet Beauty | .. 2 6 | 14 0 |
| Pink Beauty | .. 2 6 | 14 0 |
| Prize Mixed | .. 2 6 | 14 0 |

## BEDDING VIOLAS.

AMY BARR. Mauve.
BRIDAL MORN. Pale sky blue.
JOHN QUARTON. Light mauve.
LIZZIE PAUL. Yellow.
MANCHESTER BLUE. Blue.
MAGGIE MOTT. Soft mauve.

MAUVE QUEEN. Mauve.
MRS. AUDICE. White.
MRS. CHICHESTER. White flaked purple.
MRS. W. FOSTER. Light blue.
MOSELEY PERFECTION. Deep yellow.
WHITE DUCHESS. White.

Ready in March, 4d. each, 3/- per doz., 21/- 100.
All Autumn struck cuttings.

## WALLFLOWERS.

DANIELS' DOUBLE VARIETY. Tall grand spikes of double blooms, about 2 ft. high, in April and May. 4d. each, 2/- per doz., 14/- per 100.
DWARF MIXED DOUBLES. 12 inches high 4d. each, 2/- per doz., 14/- per 100.

| | per doz. | per 100 | | per doz. | per 100 |
|---|---|---|---|---|---|
| | s. d. | s. d. | | s. d. | s. d. |
| WALLFLOWER. Blood Red .. | 1 6 | 10 6 | WALLFLOWER. Eastern Queen. | | |
| „ Ellen Willmott | 1 6 | 10 6 | Chamois, changing to salmon-rose | 1 6 | 10 6 |
| „ Fire King .. | 1 6 | 10 6 | „ Primrose Monarch | 1 6 | 10 6 |
| „ Golden Monarch | 1 6 | 10 6 | „ Vulcan. Velvety crimson | 1 6 | 10 6 |
| „ Ruby Gem .. | 1 6 | 10 6 | | | |

All the above are transplanted and grown thinly, and will therefore give good results.

## VIOLETS— (Sweet-Scented).

The plants we offer are strong, well-rooted, and with good flowering crowns. If planted out in May, when flowering is over, in good soil on a shady border, they will make fine clumps for lifting in Autumn, for blooming under glass.

### SINGLE-FLOWERED VARIETIES.

ASKANIA. The freest winter flowering violet; an improvement on "Princess of Wales," the best of all single varieties.
THE CZAR. Very dark and free.

PRINCESS OF WALES. A grand variety, producing very large, beautifully formed, rich violet blue flowers.

Single-Flowered and Double-Flowered Varieties, 9d. each, 8/- per doz.   Carriage extra.

# SWEET PEAS FOR AUTUMN SOWING.

The practice of sowing Sweet Peas in Autumn has been largely revived during the last two years. The great advantages are that of early blooming and the fact that in such unfavourable Summers as the past two the well established Autumn sown plants will stand either drought or wet with equal facility. The best time for sowing outdoors is in October, in a well drained soil, but if the soil or situation is not favourable, we recommend sowing at the end of September, in boxes or pots, shaded until the seeds have germinated. The plants should be kept in frames all through the Winter, and transplanted at the proper time in Spring, when nearly a month will be gained in time of flowering over the Spring sown plants.

The following list includes most of the new varieties.

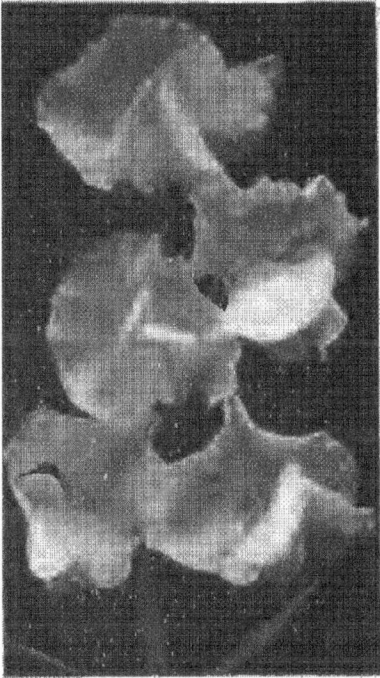

NEW SWEET PEA —MASCOTT'S WHITE.

|  | per pkt. |
|---|---|
| ANNIE IRELAND. White, edged terra-cotta .. .. .. .. .. **6d. &** | 1 0 |
| AUSTIN FREDERICK IMPROVED. Grand lavender .. .. .. .. **6d. &** | 1 0 |
| BLUE PICOTEE. Blue edge on white ground .. .. .. .. .. | 0 6 |
| BRILLIANT (new). Brilliant cherry cerise .. .. .. .. .. | 1 0 |
| CHARITY. Deep crimson .. .. **6d. &** | 1 0 |
| CONSTANCE HINTON. Pure white .. | 0 6 |
| DAISYBUD. Rich rose pink on white .. | 0 6 |
| DORIS. Cerise cherry pink .. .. | 0 6 |
| EDITH CAVELL. Rosy pink on cream .. | 0 6 |
| ELEGANCE. Soft silvery pink .. .. | 0 6 |
| GEORGE SHAWYER (new). Orange salmon | 1 0 |
| GIANT ATTRACTION. Rich shell pink .. | 0 6 |
| GLADYS. Pure lavender lilac .. .. | 0 6 |
| HAWLMARK PINK. Rose pink shaded salmon .. .. .. .. **6d. &** | 1 0 |
| JACK CORNWELL, V.C. Dark blue **6d. &** | 1 0 |
| JEAN IRELAND. Creamy buff, edged rose .. .. .. .. .. | 0 6 |
| MAJESTIC CREAM. The finest cream .. | 0 6 |
| MASCOTT'S INGMAN. Rosy carmine .. | 1 0 |
| MASCOTT'S WHITE (new). Pure white **6d. &** | 1 0 |
| MRS. ARNOLD HITCHCOCK. Salmon on cream .. .. .. .. .. | 0 6 |
| PICTURE. Rose flushed flesh pink **6d. &** | 1 0 |
| ROYAL SCOT. Brilliant scarlet .. | 0 6 |
| SENSATION. Scarlet, orange shaded .. | 0 6 |
| SPLENDOUR. Red maroon .. .. | 0 6 |
| SWEET PEAS, New Spencer Varieties. | |
| ,, 12 Fine Selected Varieties. | 4/6, |
| ,, 6 Fine Selected Varieties. | 2/6. |
| ,, Very Choice Mixed | 1/6 per oz., |
| | 6d. and 1/- per pkt. |

# HARDY ANNUALS FOR AUTUMN SOWING.

Many kinds of hardy annuals, if sown in Autumn, will withstand an ordinary Winter, and the following Spring produce blooms of a size and brilliancy that will surprise many who are only familiar with the undersized crowded specimens usually seen in our gardens during the summer months

Sow the seeds thinly, any time during September or early in October, in an open space, on rather poor soil and where they are freely exposed to air and sunshine. Here they will make good sturdy plants that will withstand the winter frosts and be in good condition for removal when required. Plant out in November or early Spring where intended to bloom—or if intended to flower where sown, the plants should be rigorously thinned out as soon as large enough to handle in Autumn.

ALYSSUM MARITIMUM. White, Dwarf. 4d. per pkt.
CANDYTUFT. Daniels' Mammoth Spiral, White. Immense spikes. 6d. per pkt.
Choicest Mixed. 4d. per pkt.
,, NEW DWARF. Choice Mixed. 6d. per pkt.
CLARKIA ELEGANS, Mixed. 6d. per pkt.
CYANUS MINOR. Choice Mixed. 4d. per pkt.
ESCHSCHOLTZIA. Extra Choice Mixed. 4d. per pkt.
GODETIA. Large-Flowered. Choice Mixed. 4d. per pkt.
LARKSPUR. Dwarf Rocket, mixed. 4d. pkt.
LUPINS. ANNUAL. Mixed. 4d. per pkt.
NIGELLA, Miss Jekyll. 6d. per pkt.

NIGELLA DAMASCENA. 4d. per pkt.
NEMOPHILA INSIGNIS. 4d. per pkt.
POPPY. Shirley Selected. Very fine, mixed. 6d. per pkt.
,, Carnation Flowered. Double mixed. 4d. per pkt.
SCABIOUS. Large-Flowered. Double Mixed. 4d. per pkt.
VIRGINIAN STOCK. Choice mixed. 4d. per pkt.
,, ,, Red. 4d. per pkt.
,, ,, White. 4d. per pkt.
VISCARIA CARDINALIS. Brilliant scarlet. 4d. per pkt.
,, OCULATA CÆRULEA. Blue. 4d. per pkt.

# SEED FOR AUTUMN GREENHOUSE SOWING.

CYCLAMEN. Daniels' Giant White. Pure white large flowers. Splendid. 1/6 per pkt.
,, VULCAN. Deep crimson. 1/6 per pkt.
,, Daniels' Giant Mixed. A magnificent strain. 1/6 and 2/6 per pkt.

CYCLAMEN. Salmon King. Large beautiful flowers of a clear salmon-pink. 1/6 per pkt.
SCHIZANTHUS WISETONENSIS. Daniels' Excelsior Strain. The finest for pot cultivation. 1/6 and 2/6 per pkt.

# BULBS FOR EARLY FLOWERING
## IN BOWLS OR GREENHOUSE.

FREESIA—NEW LARGE-FLOWERED

### DANIELS' NEW EARLY WHITE HYACINTHS.

The continually growing scarcity for several years past of White Roman Hyacinths has impelled us to seek a really efficient substitute. Offered last Autumn for the first time, our New Early White Hyacinths have received high praise from many of our customers. With early planting and attention they bloom before or during Christmas, just when a pot or bowl of flowers is most attractive, giving a display at a much lower cost than Roman Hyacinths.

DANIELS' NEW EARLY WHITE HYACINTHS. Per 100, 21/- ; per doz., 3/-.

### FREESIA REFRACTA ALBA.
#### Daniels' New Large-Flowered.

The result of years of careful selection and cultivation. The beautiful large pure white blooms, nearly twice the size of the old variety, are carried on long wiry stems, giving a most graceful effect and making it very valuable for cutting.

SELECTED BULBS.   10/0 per 100, 1/6 per dez.
ORDINARY BULBS.   6/6   ,,   1/-   ,,

## SELECTED BULBS FOR BOWLS

The following are all selected bulbs, these proving more satisfactory for Bowl Culture.

### EARLY WHITE ROMAN HYACINTHS

The most charming of all the early flowering Hyacinths. The light and graceful flower stems vary from two to four or five from each bulb according to size. The culture is of the easiest, and by early planting it may be had in bloom from November onwards.
SELECTED ROOTS. 42/- per 100 ; 5/6 per doz.
FINE ROOTS.   30/-   ,,   4/-   ,,

### MINIATURE HYACINTHS

These give spikes just the same size as our Christmas Miniature Hyacinths, but not being specially prepared do not flower until later. They form a very useful succession to the Christmas Hyacinths. Colours, same as Hyacinths in next column, 1/9 per doz. 12/6 100

### NARCISSUS

| | | per doz. |
|---|---|---|
| BARRI CONSPICUUS. White, orange eye | .. | 1 9 |
| EMPEROR. Yellow trumpet ; extra | .. | 2 9 |
| EMPRESS. Bicolour trumpet | .. | 2 9 |
| GOLDEN SPUR. Deep yellow trumpet | .. | 2 9 |
| MADAME DE GRAAFF. White trumpet | .. | 3 0 |
| SIR WATKIN. Sulphur, yellow cup | .. | 2 9 |
| WHITE LADY. White Leedsi | .. | 3 0 |

### POLYANTHUS NARCISSUS

| | | |
|---|---|---|
| PAPER WHITE. Large-flowered | .. | 3 0 |
| GRAND MONARQUE | .. | 2 6 |
| GRAND SOLIEL D'OR | .. | 2 8 |

### LACHENALIA NELSONI

LACHENALIA NELSONI. A charming little plant for early spring flowering in the greenhouse. Deep golden yellow flowers with broad fleshy leaves. Pot six in five-inch pots of turfy loam. 3/- per doz.

### IRIS RETICULATA.

IRIS RETICULATA. Beautiful little hardy Iris, intense violet blue, blotched gold ; blooms in February in the open, and does well in the cool greenhouse or in bowls.
Selected Bulbs. 3/6 per doz. } Very short
Flowering Bulbs. 2/9 per doz. } crop.

### FREESIA EXCELSIOR.

Deep cream, chrome blotch, large flower, deliciously scented. 15/- per 100 ; 2/3 per doz.

### HYACINTHS

In separate colours for bowls.    per doz.

| | | per doz. |
|---|---|---|
| BRILLIANT SCARLET | | 4 0 |
| DEEP CRIMSON | | 4 0 |
| DEEP ROSE PINK | | 4 0 |
| DELICATE PINK | | 4 0 |
| BLUSH ROSE .. | EXTRA | 4 0 |
| PURE WHITE .. | SELECTED | 4 0 |
| DEEP RICH PURPLE | BULBS. | 4 0 |
| DARK BLUE .. | | 4 0 |
| PORCELAIN BLUE | | 4 0 |
| LIGHT BLUE .. | | 4 0 |
| MAUVE OR YELLOW | | 4 0 |

NARCISSUS BARRI IN JARDINIERE.

# BULB GROWING IN BOWLS & VASES.

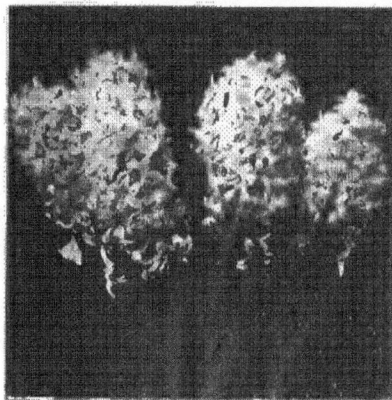

Much interest has been added to Bulb growing by the method of growing them in ornamental bowls and vases. Many reasons can be given for the popularity of this method of culture, but probably the most important one is the absolute simplicity and assurance of success, as flowers of the highest quality may be produced without any special attention beyond watering.

The material most suitable is a mixture of fibre, shell and charcoal, which we prepare specially, and is quite clean to use. Make the material nicely moist and place a fair quantity in the bottom of the bowl, then place in the bulbs and fill up round the bulbs, leaving the crowns bare at the top. Make firm, and place in a dark, cool cupboard or cellar for about a month to allow the roots to get started into growth; gradually admit light, and when the growth at the top is two or three inches high bring into the room, and place in a light position. Nothing more will now be needed than to keep an eye on them for water (no set rule can be given for watering, it will depend chiefly on the temperature of the room), but don't allow them to get too dry or too wet.

## CHRISTMAS HYACINTHS.

These beautiful Hyacinths, introduced by us, have fully merited the confidence with which we recommended them; carefully grown, they will flower well in advance of the ordinary Hyacinths. We also offer these prepared Hyacinths in a smaller size suitable for bowl culture, which if planted early will give a fine display at the festive season.

| | | Extra Large Bulbs | | Miniature Bulbs | |
|---|---|---|---|---|---|
| | | s. d. | s. d. | | s. d. |
| GENERAL PELISSIER. Crimson, very fine | per doz. | 7 6 | each 0 8 | per doz. | 2 6 |
| EXCELSIOR (new). Lovely pink, the earliest to bloom | ,, | 7 6 | ,, 0 8 | ,, | 3 6 |
| MORENO. Deep rosy blush | ,, | 6 6 | ,, 0 7 | ,, | 2 6 |
| VICTOR EMMANUEL. Beautiful rose pink | ,, | 8 6 | ,, 0 9 | ,, | 3 3 |
| CAPTAIN BOYTON. Porcelain blue | ,, | 7 6 | ,, 0 8 | ,, | 2 9 |
| QUEEN OF THE BLUES. Charming light blue, large truss | ,, | 8 6 | ,, 0 9 | ,, | 3 0 |
| L'INNOCENCE. Pure white, the best white | ,, | 6 6 | ,, 0 7 | ,, | 2 6 |
| YELLOW HAMMER. Pure yellow | ,, | 7 6 | ,, 0 8 | ,, | 3 0 |

# TULIPS FOR EARLY FLOWERING.

## SINGLE TULIPS.

The Van Thols are particularly valuable for early forcing, and by potting early in September, and gently forcing they may be had well in bloom by Christmas. Pot three or five in a pot of five or six inches diameter. Soil and after treatment as recommended for Hyacinths.

DUC VAN THOL, COCHINEAL. Brilliant colour, the largest and earliest of the Van Thols. 14/- per 100, 2,- per doz.

DUC VAN THOL, SCARLET. Splendid. 12 6 per 100, 1/10 per doz.

DUC VAN THOL, YELLOW. Clear yellow. 12/- per 100, 1/9 per doz.

DUO VAN THOL, WHITE. 12,6 per 100, 1/10 per doz.

## DOUBLE TULIPS.

The large handsome flowers of the double tulips are most attractive when gently forced, and the colours are much more delicate.

COURONNE DES ROSES. Lovely deep rose and white. 18/6 per 100, 2/8 per doz.

LUCRETIA. Bright rose, charming. 11/- per 100, 1/8 per doz.

MURILLO. Delicate rose with white. 8/6 per 100, 1/3 per doz.

PARMESIANO. Beautiful pale rose. 21/- per 100, 3/- per doz.

PURITY. The finest pure white. 16,6 per 100, 2,4 per doz.

SAFRANO. Most beautiful yellow and salmon. 12/- per 100, 1/9 per doz.

## BOWLS OF BULBS COMPLETE.

We can supply Bowls, planted with New Early White Hyacinths, which if secured early in the season will bloom during December. Bowl planted complete 5/6 each. Packing and postage 1/6 extra.

We can supply similar Bowls planted with large Dutch Hyacinths—Pink, White or Blue, which will flower in February and March or earlier if needed, at the same prices as quoted.

Also Bowls planted with Daffodils, Paper White Narcissus, White or Purple Crocus, at 4/- and 5/6 each. Packing and postage 1/6.

## BULB BOWLS.

We have a very fine assortment of bowls of various designs suitable for Bulb Growing, and if our customers will leave the choice to us and give an idea of the price they wish to pay, we will endeavour to select something that will be sure to please. The colours are chiefly green, but we can also supply pink and blue, and also some very nice ivory coloured. Prices ranging from 2/6 to 15/-.

JAPANESE BOWLS. In beautiful designs for Bulb Growing or for use as Rose bowls. Prices 15/- to 25/-, packing and carriage extra.

## SHELL FIBRE.

This is a preparation of Fibre, with Shell and Charcoal in the proper proportions, and has become very popular with our customers. We have very carefully tested it and can recommend it with every confidence. Per peck 2,6, carriage free; per bushel 7,-, bag and carriage 2,-.

# DANIELS' HYACINTH COLLECTIONS.

## SUPERB EXHIBITION VARIETIES.

We have very great pleasure in recommending the following superb Collections for pot culture or exhibition. These Collections contain all the most splendid varieties, and include the most beautiful shades of colour. Toward the end of the planting season it may become necessary to make some minor alterations in these Collections.

†Charles Dickens, Improved, deep porcelain
**†City of Haarlem, pure yellow
* Correggio, pure white
†Etna, rosy carmine
†Garibaldi, brilliant crimson
**†Gounod, lavender
Ivanhoe, rich dark blue
†Jacques, deep rose pink
**†King Alfred, lilac blue
**†King of the Scarlets, deep scarlet
†La Grandesse, pure white
†Louis Pasteur, deep porcelain

*†Marconi, bright rose
Mauve Queen, clear mauve
Perfection Rose, light rose pink
Primrose Perfection, primrose yellow
**†Prince Henry, pale rose
**†Princess Juliana, blush white
* Queen of the Pinks, rose pink
**†Queen of the Whites, pure white
†Simplicity, pure white
**†Victory, crimson scarlet
†Washington Irving, light blue, tinted lilac
**†Westminster, purple blue

24 Exhibition Hyacinths as named above 14/-. Carriage Free.
18 „ „ as marked † 10/6. „ „
12 „ „ as marked * 7/6. „ „

# DANIELS' BEDDING HYACINTHS.

## IN DISTINCT AND BEAUTIFUL COLOURS. ALL SELECTED BULBS.

These fine selected and distinct colours give a most charming effect when planted in small beds of a single shade, or two well matched varieties in a bed will give an even more lovely display. The bulbs require planting about six inches apart to give the best effect. All the varieties offered below are single-flowered, these being preferable to the double for bedding.

| | Per 100. | Per doz. | | Per 100. | Per doz. |
|---|---|---|---|---|---|
| Daniels' Brilliant Scarlet | 27/- | 3/6 | Daniels' Deep Rich Purple | 27/- | 3/6 |
| Daniels' Rich Crimson | 27/- | 3/6 | Daniels' Dark Blue | 27/- | 3/6 |
| Daniels' Deep Rose Pink | 27/- | 3/6 | Daniels' Porcelain Blue | 27/- | 3/6 |
| Daniels' Delicate Pink | 27/- | 3/6 | Daniels' Light Blue | 27/- | 3/6 |
| Daniels' Blush Rose | 27/- | 3/6 | Daniels' Mauve | 27/- | 3/6 |
| Daniels' Pure White | 27/- | 3/6 | Daniels' Pure Yellow | 27/- | 3/6 |
| Mixed, extra choice, from above splendid sorts | | | .. .. | per 100, 24/- ; per doz. 3/3 | |

# MIXED HYACINTHS
## for BEDS and BORDERS.

The following list contains various shades of the colours named.

Distinct colours are much the best for bedding, but for filling odd places and for growing on borders the following are quite satisfactory. To give any effect they should not be planted in clumps of less than five or six. If left in the ground the bulbs will give a display the succeeding year, although the spikes will naturally be a good deal smaller.

## SINGLE-FLOWERED.

| | per 100. | per doz. | | | per 100. | per doz. |
|---|---|---|---|---|---|---|
| Red, rose, carmine and crimson | 16/6 | 2/4 | Yellow shades | .. | 16/6 | 2/4 |
| White, pure and slightly tinted | 16/6 | 2/4 | MIXED, all colours | | 14/- | 2/- |
| Light and dark blue and purple | 16/6 | 2/4 | | | | |

### OPEN GROUND CULTIVATION.

Hyacinths succeed in any good well drained garden soil, but they like a liberal feeding, and good rich soil is always preferable if available. The soil should be forked up to a depth of 18 inches and a good quantity of well-decayed manure worked in, and if the ground is at all heavy a dressing of coarse sand will be beneficial. The bulbs should be planted in October, being placed about 6 inches apart, and the crowns buried to a depth of not less than 4 inches. If the beds are in an exposed position, they should be covered with bracken or spruce boughs, but otherwise nothing further will be needed, and a good display of bloom may confidently be expected.

# DANIELS' SUPERB HYACINTHS.

*Double Flowered Varieties are denoted
by the letter (D)*

## SECTION I.

**SCARLET, CRIMSON, ROSE, DEEP AND PALE PINK.**

**CARDINAL WISEMAN.** Large compact spike, deep pink, one of the finest for pot culture. 5/6 per doz., 6d. each.

**CHESTNUT-FLOWER (D).** Bright rose, large spike and flower. 5/6 per doz., 6d. each.

**ETNA.** Deep rosy carmine, large fleshy bells, forming a large spike on a strong stem. Late flowering. 7/- per doz., 8d. each.

**GARIBALDI.** Glowing crimson, magnificent spike, one of the finest exhibition varieties. 5/6 per doz., 6d. each.

**GENERAL PELISSIER.** Deep crimson, fine spike, early flowering. 5/6 per doz., 6d. each.

**JACQUES.** Deep rosy pink with splendid spike of large beautiful flowers. Highly recommended. 6/- per doz., 7d. each.

**KING OF THE BELGIANS.** Dark bright crimson, large compact spike; one of the best. 5/6 per doz., 6d. each.

**KING OF THE SCARLETS (New).** Very fine deep scarlet, good spike. 7/- doz., 8d. each.

**KOH-I-NOOR (D).** Large, semi-double flowers of a beautiful salmon-rose colour, borne on a splendid spike. 6/- per doz., 7d. each.

**LADY DERBY.** Lovely pale rose, magnificent spike of large well-shaped flowers, one of the finest varieties yet raised. 5/6 per doz., 6d. each.

**MARCONI.** Bright rose pink, broad truss on a stiff compact stem. 7/- per doz., 8d. each.

**MORENO.** Deep rosy blush with a fine long spike. Extra fine. 5/6 per doz., 6d. each.

**ORANGE BRILLIANT (New).** Brilliant orange scarlet, the finest in this particular shade. This variety only develops a small bulb. 7/- per doz., 8d. each.

**ORNEMENT ROSE.** Soft flesh-pink, very large, broad spike. 5/6 per doz., 6d. each.

**PERFECTION ROSE (New).** Large truss, light rose pink. 6/- per doz., 7d. each.

**PINK PERFECTION.** Bright rosy pink, large spike; splendidly effective. 5/6 doz., 6d. each.

**PRESIDENT ROOSEVELT (D).** Large, broad, compact spike, colour a lovely shade of pink. 6/- per doz., 7d. each.

**PRINCE HENRY (New).** Closely placed bells on fine spike, lovely pale rose. 7/- doz., 8d. ea.

**PRINCE OF ORANGE (D).** Pale red with carmine stripe, semi-double flower, fine truss. 6/- per doz., 7d. each.

**QUEEN OF THE PINKS.** Beautiful bright rosy pink, grand spike. A splendid exhibition variety. 6/- per doz., 7d. each.

**ROSE A MERVEILLE.** Beautiful bright rose, fine spike, keeping its colour splendidly. 5/6 per doz. 6d. each.

**VICTORY (New).** Undoubtedly the very finest of the scarlet varieties, the long spike of closely-placed bells giving a brilliant glowing effect. 6/- per doz., 7d. each.

HYACINTH—VICTORY.

# DANIELS' SPECIAL DECORATIVE COLLECTIONS.

Arranged for Conservatory and Greenhouse Decoration.

*Arentine Arendsen, snow white
*Enchantress, bright porcelain
*General Pelissier, deep crimson
*Grandeur a Merveille, delicate blush
Grand Monarque, porcelain blue
King of the Belgians, crimson
*King of the Blues, dark blue
*King of the Yellows, bright yellow
Lady Derby, pale rose

*L'Innocence, pure white
Lord Balfour, mauve striped
Madame Van der Hoop, white
*Menelik, intense blue-black
*Moreno, deep rosy blush
*Pink Perfection, rosy pink
*Queen of the Blues, light blue
*Rose a Merveille, bright rose
Schotel, pale blue

The above 18 choice varieties for 9/-
12 choice varieties as marked * 6/-
3 each of the above 18 varieties for 25/-
3 each of the 12 varieties as marked * 17/6
} Carriage Free.

## COLLECTIONS OF GOOD VARIETIES FOR POT CULTURE.

Our Own Selection only.

18 in 18 varieties 6/6, 12 in 12 distinct varieties 4/6, 6 in 6 fine varieties 2/6, Carriage Free.

# DANIELS' SUPERB HYACINTHS.

The following list comprises the most distinct and beautiful of the white and tinted varieties, and customers will find it less difficult to make their selection than from the old lengthy lists of almost identical varieties.

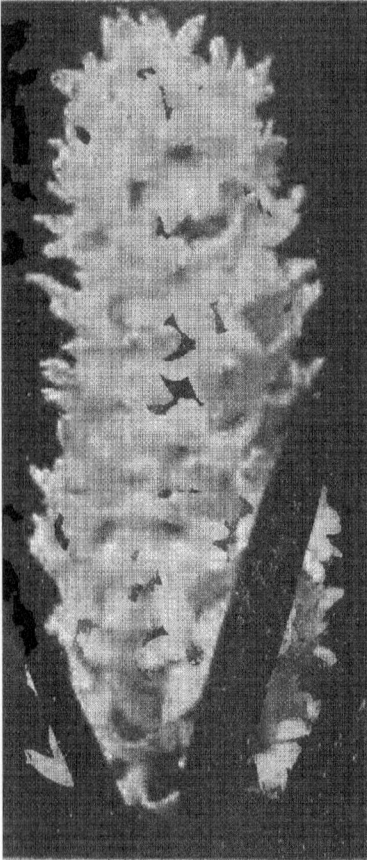

HYACINTH—CORREGGIO.

*Double Flowered Varieties are denoted by the letter (D).*

## SECTION II.

PURE WHITE, BLUSH WHITE AND CREAMY WHITE.

ANGENIS CHRISTINA. Pure white, very fine ; a beautiful variety for pots or vases, and makes a very fine bedder. 5/6 per doz., 6d. each.

ARENTINE ARENDSEN (New). One of the very finest of recent introduction, the large flowers with broad open petals are of an intense snow white ; the large well-rounded spike is unusually well filled. This will become one of the leading varieties. 6/- per doz., 7d. each.'

BARONESS VAN THUYLL. An old well-known sort with a good spike of pure white flowers, very early flowering, and most useful for forcing. 6/- per doz., 7d. each.

CORREGGIO (New). Fine spike of large bells of the purest white. One of the finest for exhibition. 7/- per doz., 8d. each.

EDISON (D). Ivory white, beautiful spike, large bells. 6/- per doz., 7d. each.

GRANDEUR A MERVEILLE. Delicate blush white, large handsome truss of fine large flowers. 5/6 per doz., 6d. each.

LA GRANDESSE. Pure white, very large bells and fine spike. A superb variety for pot culture, and undoubtedly one of the finest single pure white hyacinths ever raised. 6/- per doz., 7d. ea.

LA TOUR D'AUVERGNE (D). By far the earliest of all double white hyacinths. The blooms, which are produced in long handsome spikes, are of the purest white, and in consequence of its earliness it is one of the very best for forcing. 6/- per doz., 7d. each.

L'INNOCENCE. Pure white, very fine trusses. A good exhibition variety. 5/6 per doz., 6d. each.

MADAME VAN DER HOOP. A good pure white with large bells and truss, medium early, makes a good pot hyacinth. 5/6 per doz., 6d. each.

MR. PLIMSOLL. Beautiful ivory white, fine large spike, good forcing variety. 5/6 doz., 6d. ea.

PRINCESS JULIANA (new). Enormous bells of creamy white, very fine truss. 9/- doz., 10d. each.

QUEEN OF THE WHITES (now). Fine exhibition variety, broad pure white spike. 8/- doz., 9d. ea.

SIMPLICITY (now). Very fine pure white, good for pots. 6/- per doz., 7d. each.

SNOW QUEEN (now). Large broad snow-white truss. 6/- per doz., 7d. each.

TUBEROSE (D). Large spikes of pure white flowers, with long pointed petals resembling Tuberose blooms in form, deliciously fragrant ; a very fine variety. 6/- per doz., 7d. each.

WHITE LADY (New). A lovely pure white, with large broad compact spike of beautiful flowers ; a quite new and charming introduction. 7/- per doz., 8d. each.

---

## CULTIVATION OF HYACINTHS IN POTS.

Long experience has taught us that the best possible mixture for growing Hyacinths in pots consists of about one-third each of good fibrous loam and well-decayed cow-dung, and the remainder of about equal parts of coarse gritty sand and leaf-mould. Mix at least a month before potting, keeping the mixture in a cool shed, or where it is not exposed. Five-inch pots are generally preferred. After potting select a cool, sheltered position out of doors, and place them on a layer of coal ashes, cocoa-nut fibre, tan, or any similar light material, covering the pots with the same to the depth of five or six inches, and placing over them a few boards or slates to keep off excessive rains.

In about five weeks they will be found to be nicely rooted, and those required for earliest blooming may be removed to the cool frame preparatory to forcing, or be brought forward as required in succession for later blooming ; all should, however, be removed before they push their way through, and be gradually introduced to light and air, keeping them in a cool shady position till the leaves have fairly acquired their proper greenness. It may be here remarked, that a slow and steady development is at all times preferable for the production of really fine spikes of bloom. Although the Hyacinth forces well, it should never be subjected to heat before the bulbs are well rooted or the flower spike is liable to shrivel, and they should never be placed in a higher temperature than sixty-five or seventy degrees, from which they should be removed to a cooler atmosphere as soon as the flower-spikes are formed above the surface.

# DANIELS' SUPERB HYACINTHS.

The following section contains the names of the choicest possible selection of all shades of blue, from the very palest silvery lilac, through shades of porcelain, to deep blue and almost black.

*Double Flowered Varieties are denoted by the letter (D).*

HYACINTH—IVANHOE.

**WESTMINSTER (New).** Extra fine spike of very large flowers. The colour is a deep purple blue with a distinctly marked white centre; very fine variety. 6/- doz., 7d. each.

## SECTION III.

### PALE TO DARK BLUE, LILAC AND PURPLE.

**CAPTAIN BOYTON.** Dark porcelain blue with splendid truss of large handsome bells. A first-rate variety. 6/- per doz., 7d. each.

**CHAS. DICKENS, IMPROVED.** A fine early variety. Flowers deep porcelain blue, splendid truss. 7/- per doz., 8d. each.

**DELICATA (D).** Very good double-flowered light blue, with a really fine spike. 7/- per doz., 8d. each.

**ELECTRA.** Very pale blue, enormous truss of very large flowers. 6/- per doz., 7d. each.

**ENCHANTRESS.** Bright porcelain blue. Immense spikes of large bells. A very fine variety. For exhibition. 6/- per doz., 7d. each.

**GARRICK (D).** Light bright blue, one of the finest double varieties. 6/- per doz., 7d. each.

**GOUNOD (New).** Dark lavender blue, large splendidly shaped spike. 8/- doz., 9d. each.

**GRAND LILAS.** Beautiful silvery lilac, large bells, extra fine spike. 6/- per doz., 7d. each.

**GRAND MONARQUE.** Very fine porcelain blue, one of the best varieties. 5/6 per doz., 6d. each.

**IVANHOE.** Rich glittering dark blue, almost black, fine spikes, splendid exhibition variety. 7/- per doz., 8d. each.

**JOHAN.** Pale greyish-blue. Immense spikes of bloom. A quite distinct and splendid variety that can be highly recommended. 5/6 per doz., 6d. each.

**KING ALFRED (New).** Splendid spikes of large bells, lilac shaded plum-purple, well lighted up in centre. A fine exhibition flower. 6/- per doz., 7d. each.

**KING OF THE BLUES.** The finest dark blue, long spike, closely filled with flowers, good both for pots or glasses. 5/6 per doz., 6d. each.

**LAVENDER BELLE (New).** Beautiful lavender, shaded lilac. 6/- per doz., 7d. each.

**LA VICTOIRE (D).** Fine spike of bright violet. 6/- per doz., 7d. each.

**LOUIS PASTEUR (New).** Deep porcelain, the largest truss of any Hyacinth. 8/- doz., 9d. each.

**MENELIK.** Intensely deep black blue, very large spike; splendid variety. This fine hyacinth is quite unique in colour. 6/- per doz., 7d. each.

**PEARL BRILLIANT.** Pale blue, the tubes a brilliant blue, a variety of pleasing effect and first-class for exhibition. 6/- per doz., 7d. each.

**QUEEN OF THE BLUES.** Very beautiful large light blue truss of fine flowers, a very good and useful variety for early blooming. 6/- per doz., 7d. each.

**SCHOTEL.** Pale blue, very fine spike, a first-class exhibition variety. 6/- doz., 7d. each.

**WASHINGTON IRVING (New)** Beautiful light blue, tinted lilac.

## SECTION IV.

### CLARET AND MAUVE—A BEAUTIFUL CLASS.

**DISTINCTION.** A quite distinct and beautiful variety, the colour being a fine deep reddish violet mauve. 6/- per doz., 7d. each.

**L'ESPERANCE.** A splendid variety, unusual shade of rich claret. 6/- per doz., 7d. each.

**LORD BALFOUR.** Mauve striped, large flowers and grand spike, quite distinct. 5/6 per doz., 6d. each.

**MAUVE QUEEN.** Fine truss of flowers, colour of a really clear mauve. 6/- per doz., 7d. each.

**SIR WILLIAM MANSFIELD.** Beautiful bright purplish mauve, fine long spikes. 5/6 per doz., 6d. each,

## SECTION V.

### APRICOT, YELLOW, AND PRIMROSE.

**BUFF BEAUTY.** Buff orange, unusual shade, long truss on strong stem. 6/- per doz., 7d. ea.

**CITY OF HAARLEM.** By far the finest of the golden yellows, enormous spike of large flowers. 7/- per doz., 8d. each.

**KING OF THE YELLOWS.** Bright yellow, fine large bells and spike, good for pots, rather a small bulb. 5/6 per doz., 6d. each.

**PRIMROSE PERFECTION.** Beautiful primrose yellow, splendid spike. First Class Certificate. 6/- per doz., 7d. each.

**YELLOW HAMMER.** Beautiful golden yellow; long truss of flowers, one of the finest for forcing or decoration in bowls. 6/- per doz., 7d. each.

# DANIELS' CHOICE SINGLE TULIPS.

Carefully selected varieties of these beautiful flowers, planted in beds with due regard to arrangement of the colours, produce some of the most charming and fascinating effects.

## EARLIEST FLOWERING.

**DUC VAN THOL, COCHINEAL.** Brilliant colour, the largest and earliest Van Thol. 14/- per 100, 2/- per doz.

**DUC VAN THOL, SCARLET.** Brilliant colour. Height 7 ins. 12/6 per 100, 1/10 per doz.

**DUC VAN THOL, YELLOW.** Fine colour yellow. Height 7 ins. 12/- per 100, 1/9 per doz.

**DUC VAN THOL, WHITE.** Pure white. Height 7 ins. 12/6 per 100, 1/10 per doz.

## GENERAL LIST.

Those marked with an asterisk (*) bloom about seven or eight days later than those not marked.

**\*ARTUS.** Scarlet, a fine variety. Height 8 ins. 10/- per 100, 1/6 per doz.

**BELLE ALLIANCE.** Brilliant scarlet. Height 7 ins. 15/- per 100, 2/3 per doz.

**CALYPSO.** A splendid bedding variety, the colour is a lovely shade of primrose yellow, quite distinct. Height 12 ins. 10/- per 100, 1/6 per doz.

**CANARY BIRD.** Beautiful canary yellow. Height 10 in. 11/- per 100, 1/8 per doz.

**\*CARDINAL'S HAT.** Crimson scarlet with gold edge. Ht. 8 ins. 11/- per 100, 1/8 per doz.

**CHRYSOLORA.** Fine pure yellow ; capital bedder. Height 9 ins. 11/- per 100, 1/8 doz.

**\*COLEUR CARDINAL.** Velvety crimson, shaded plum, intense colouring ; large and substantial flower. Height 12 ins. 18/6 per 100, 2/8 per doz.

**COLEUR PONCEAU.** Rosy crimson and white, good bedder. 12/6 per 100, 1/10 per doz.

**COTTAGE MAID.** Dark rose, shaded white ; one of the best bedders. Height 7 ins. 14/- per 100, 2/- per doz.

**\*DUCHESSE DE PARMA.** Brown red with bold margin of deep yellow. Height 10 ins. 14/- per 100, 2/- per doz.

**\*DUSART.** Deep vermilion scarlet, a splendid bedder. Height 8 ins. 12/- 100, 1/9 doz.

**FRED MOORE.** A fine shade of deep orange. A grand variety for early flowering. Height 10 ins. 10/- per 100, 1/6 per doz.

**\*GOLDFINCH.** Very large pure yellow flowers ; good bedder or forcer. Height 10 ins. 12/6 per 100, 1/10 per doz.

**GOLDEN QUEEN.** Pure yellow, large flowers with substantial petals, lasting long in condition, a splendid bedder. 14/- per 100, 2/- per doz.

**GRAND DUKE** (Keizerskroon). Crimson-scarlet, deep yellow edge ; a tall handsome tulip. Height 12 ins. 12/- per 100, 1/9 per doz.

**\*JOOST VAN VONDEL.** Large handsome flower, cherry-red, feathered white. 11/- per 100, 1/8 per doz.

**LA GRANDEUR.** One of the most beautiful, very brilliant scarlet, large flower. Height 10 ins. 17/- per 100, 2/6 per doz.

**LA REINE.** White, suffused delicate pink, very pretty. Height 9 ins. 8/- per 100, 1/3 per doz.

**\*LADY BOREEL.** Snowy-white flowers of perfect shape ; the finest white bedding Tulip. Height 12 ins. 15/- per 100, 2/3 per doz.

**\*L'IMMACULE.** Splendid pure white. Height 9 ins. 12/6 per 100, 1/10 per doz.

**MAES.** Dazzling scarlet, grand colour. Flower of immense size. Height 12 ins. 14/- per 100, 2/- per doz.

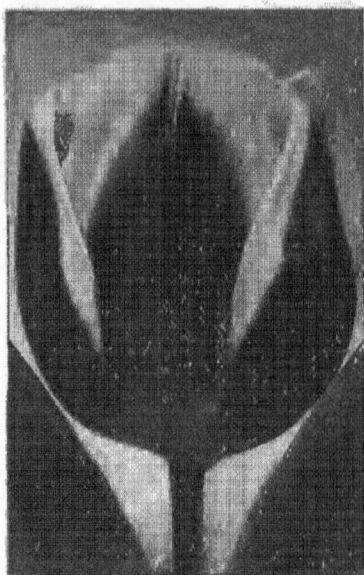

SINGLE TULIP—GRAND DUKE.

**MON TRESOR.** Immense golden yellow. Height 11 ins. 12/6 per 100, 1/10 per doz.

**\*OPHIR D'OR.** A lovely pure yellow, one of the best bedding varieties, large flower. 12/- per 100, 1/9 per doz.

**PELICAN** (New). A great improvement on the old White Hawk. A grand pure white with pointed petals. Height 10 ins. 22/6 per 100, 3/3 per doz.

**\*PINK BEAUTY.** Delicate creamy white, deeply edged with deep rose. Splendidly effective. Height 9 ins. 21/- 100, 3/- doz.

**\*PRIMROSE QUEEN.** Beautiful and distinct variety, canary yellow, stained primrose. Height 10 ins. 8/6 per 100, 1/3 per doz.

**\*PRINCE OF AUSTRIA.** Bright orange-scarlet, shading off to orange-buff, sweet-scented. Height 10 ins. 12/6 per 100, 1/10 per doz.

**PROSPERITY** (New). A most beautiful variety ; the flowers are of a lovely shade of rose pink. Height 12 ins. 14/- per 100, 2/- per doz.

**QUEEN OF THE PINKS.** Rose and white, splendid bedder, continues in bloom a long time. Height 9 ins. 15/- per 100, 2/3 per doz.

**QUEEN OF VIOLETS.** Violet, white base, and distinct. Height 10 ins. 12/6 per 100, 1/10 per doz.

**RED ADMIRAL** (New). A splendid and effective bedding variety of recent introduction ; immense flowers of a rich orange scarlet. Height 10 ins. 20/- per 100, 2/9 per doz.

**\*ROSE GRISDELIN.** Dark rose flushed white, very lovely. Height 7 ins. 10/- per 100, 1/6 per doz.

**\*ROYAL STANDARD.** Crimson, striped and feathered golden yellow, very showy. 16/6 per 100, 2/4 per doz.

# DANIELS' CHOICE SINGLE TULIPS.

**ROSE LUISANTE.** Beautiful bright rose, paler edge, the finest rose-coloured tulip. Height 9 ins. 22/- per 100, 3/- per doz.

**SIR THOMAS LIPTON.** Brilliant scarlet, magnificent flower. Height 10 ins. 21/- per 100, 8/- per doz.

**\*SPARKLER (New).** Large, bold, perfectly shaped, flower of brilliant glowing scarlet. Height 12 ins. 10/- per 100, 1/8 per doz.

**STANLEY.** Rich glowing rose. Height 9 ins. 12/6 per 100, 1/10 per doz.

**\*THOMAS MOORE.** Orange-buff; one of the most distinct and effective varieties. Height 9 ins. 10/- per 100, 1/6 per doz.

**VAN DER NEER.** Rich purple violet, large flower. Height 10 ins. 17/6 per 100, 2/6 doz.

**WATERLOO.** Brilliant crimson scarlet; very fine. Height 8 ins. 15/- per 100, 2/8 per doz.

**WHITE HAWK.** Pure white, pointed petals. Height 11 ins. 17/6 per 100, 2/6 per doz.

**\*WHITE SWAN (true).** Large egg-shaped, pure white flowers; splendid. Height 12 ins. 15/- per 100, 2/3 per doz.

**WOUVERMAN.** Dark claret purple, Height 9 ins. 12/- per 100, 1/9 per doz.

**\*YELLOW PRINCE.** Bright golden yellow, sweet-scented, splendid bedder. Height 9 ins. 11/- per 100, 1/8 per doz.

## Choice Named Single Tulips in Collections. (Our Selection).

| | |
|---|---|
| **50 in 5 choice varieties, 6/6** | **24 in 8 choice varieties, 3/6** |
| **36 in 9 choice varieties, 5/-** | **12 in 4 choice varieties, 2/-** |

# DANIELS' SINGLE MIXED TULIPS.

**Very Choice Sorts, per 100, 10/-; per doz., 1/6   Good Showy Kinds, per 100, 6/6; per doz., 1/-**

# DANIELS' CHOICE DOUBLE TULIPS.

The best Varieties selected especially for their fine double flowers and perfection of colour.

**BLEU CELESTE.** Soft violet blue, large flowers. Height 15 ins. 15/- per 100, 2/3 per doz.

**BOULE DE NEIGE.** Pure white, very beautiful. Height 9 ins. 15/- per 100, 2/3 per doz.

**COURONNE DES ROSES.** Dark rose and white. Height 9 ins. 18/6 per 100, 2/8 doz.

**COURONNE D'OR.** Yellow, shaded orange, large flower on erect stems; fine bedding or pot Tulip. Height 10 ins. 12/6 per 100, 1/10 per doz.

**EL TOREADOR.** Bright orange scarlet with buff yellow border, most conspicuous, very fine for pots. 25/- per 100, 3/6 per doz.

**FIRE KING (New).** Brilliant orange-red, very showy. Height 11 ins. 21/- per 100, 3/- doz.

**GLORIA SOLIS.** Crimson scarlet with golden stripes. Height 9 ins. 14/- per 100, 2/- doz.

**INNOCENCE.** Early flowering, pure white, fine for pot culture. Height 8 ins. 18/6 per 100, 2/4 per doz.

**LA GRANDESSE.** Lovely pink, slightly shaded white, very double flower. Height 12 ins. 14/- per 100. 2/- per doz.

**LEONARDO DA VINCI.** Orange with paler edge. Height 10 ins. 16/6 per 100, 2/4 doz.

**LINA.** Deep rose, shaded white at the base. Height 10 ins. 12/- per 100, 1/9 per doz.

**LE MATADOR.** A magnificent variety, large double flowers of glowing scarlet, with purple bloom on the outside petals. Height 11 ins. 14/- per 100, 2/- per doz.

**LUCRETIA.** Deep bright rose, very charming. Height 12 ins. 11/- per 100, 1/8 per doz.

**MURILLO.** A most beautiful variety either for pots or bedding; lovely delicate rose and white. Height 10 ins. 8/6 per 100, 1/3 doz.

**PARMESIANO.** Large splendid flowers, beautiful pale rose, splendid bedder or for pots. Height 10 ins. 21/- per 100, 3/- per doz.

**PURITY (New).** This charming introduction is the finest pure white double tulip yet raised, and will give equally satisfactory results both for pot culture or for bedding. Height 10 ins. 16/6 per 100, 2/4 per doz.

**REVE D'OR (Golden Dream) (New).** Large flowers, pure yellow. Height 11 ins. 21/- per 100, 3/- per doz.

**ROSE BLANCHE.** Pure white, a splendid variety. Height 7 ins. 16/6 per 100, 2/4 per doz.

**RUBRA MAXIMA.** The finest double deep red. Height 8 ins. 12/- per 100, 1/9 doz.

**SAFRANO.** Large flowers of a beautiful yellow and salmon shade, resembling a Marechal Niel Rose; one of the finest. Height 9 ins. 12/- per 100, 1/9 per doz.

**SALVATOR ROSA.** Deep rose slightly flushed white. The best pink double tulip for early flowering. 14/- per 100, 2/- per doz.

**VUURBAAK (Fire Dome).** The finest of all the double scarlet tulips. Immense flowers developing an orange shading when fully expanded. 21/- per 100, 3/- per doz.

**YELLOW ROSE.** Pure yellow, sweet-scented flowers, lovely colour. Height 9 ins. 14/- per 100, 2/- per doz.

## Choice Named Double Tulips in Collections. (Our Selection).

| | |
|---|---|
| **50 in 5 choice varieties, 7/6.** | **24 in 6 choice varieties, 3/6.** |
| **36 in 6 choice varieties, 5/6.** | **12 in 4 choice varieties, 2/-.** |

**DANIELS' DOUBLE MIXED TULIPS, Very Choice Sorts ... per 100, 10/-; per doz. 1/6**

# DANIELS' "MATCHLESS" BEDDING TULIPS.

| | per 100. | per doz. | | per 100 | per doz. |
|---|---|---|---|---|---|
| Single Pure White .. | 15/- | 2/3 | Double Pure White .. | 17/- | 2/6 |
| Single Rose Pink .. | 18/6 | 2/8 | Double Deep Pink .. | 17/- | 2/6 |
| Single Pure Yellow .. | 14/- | 2/- | Double Scarlet and Gold | 17/- | 2/6 |
| Single Scarlet .. | 17/- | 2/6 | Double Bright Violet .. | 17/- | 2/6 |
| Single Rose and White .. | 17/- | 2/6 | Double Scarlet .. | 17/- | 2/6 |

# DARWIN TULIPS—May Flowering.

This is one of the most magnificent classes of all garden bulbs for late Spring blooming. The flowers, which are borne on tall stately stems twenty-four to thirty inches in height, are of a great range of charming colours. from white to the deepest maroon black.

One of the most valuable qualifications of Darwin Tulips is their great value in bridging the interval between the earlier Spring flowers and the Summer bedding plants, a time when there is little else to admire in the garden. The flowers are also, by reason of their thick, fleshy textured petals, very lasting, and a fine bed of these grand tulips seen in brilliant Spring sunshine is a sight not soon forgotten.

**ISIS.** Fiery crimson scarlet, very large flowers of striking beauty. One of the best of the new varieties. 14/- per 100, 2/- per doz.

**KING HAROLD.** Rich glowing ruby crimson, enormous flowers. 17/- per 100. 2/6 per doz.

**LA COQUETTE.** Rich rose, silvery rose margin. 12/- per 100, 1/9 per doz.

**LA TULIPE NOIRE** (The Black Tulip). Darkest colour of all Tulips, intense maroon black with lustrous sheen. 15/- per 100, 2/3 per doz.

**LOVELINESS.** Soft rosy carmine, a very lovely Tulip, deepening in colour as the bloom ages. 10/- per 100, 1/6 per doz.

**MAY QUEEN.** Large open flower, salmon rose shaded lilac. 12/6 per 100, 1/10 per doz.

**MARGARET.** Pale silvery rose, flushed white, slightly deeper in shade inside; a beautiful flower of delicate colouring. 11/- per 100, 1/8 per doz.

**MASSACHUSETTS.** Vivid pink, tinted purple outside, white base. A striking and effective variety. 17/- per 100, 2/6 per doz.

**MAUVE CLAIRE.** Beautiful lilac mauve. 14/- per 100, 2/- per doz.

**MILLET.** Crimson maroon, extra fine. 14 - per 100, 2/- per doz.

**MR. FARNCOMB SANDERS.** Rich rose scarlet, centre white. 16/6 per 100, 2/4 per doz.

**MRS. KRELAGE.** Large soft rose, margined blush. 16/6 per 100, 2/4 per doz.

**NAUTICUS.** Cerise rose, violet base, shaded bronze, very large. 16/6 per 100, 2/4 per doz.

**NORA WARE.** Silvery lilac, darker toward the centre. 11/- per 100, 1/8 per doz.

**OUIDA.** Bright crimson-scarlet, one of the best for bedding. 16/6 per 100, 2/4 per doz.

**PAINTED LADY.** Soft milky white, centre of the petals lightly flushed heliotrope. Dark stem, good bedder. 14/- per 100, 2/- per doz.

**PHYLLIS.** Lovely delicate shade, white flushed rosy lilac. 14/- per 100, 2/- per doz.

**PRIDE OF HAARLEM.** A splendid flower of great size; bright rose suffused with purple. 12/6 per 100, 1/10 per doz.

**PROFESSOR F. DARWIN.** Bright cerise scarlet, blue base; one of the most distinct and brilliant. 16/6 per 100, 2/4 per doz.

**PSYCHE.** Soft rose, with interior of petals silvery rose. 14/- per 100, 2/- per doz.

**QUEEN OF ROSES.** An erect flowering variety; lovely shade of purplish rose with blush rose edge. 15/- per 100, 2/3 per doz.

**RUBY QUEEN.** Brilliant ruby red, large broad petalled flower. 12/- per 100, 1/9 per doz.

**SALMON KING.** Bright glowing salmon rose. 15/- per 100, 2/3 per doz.

**SCARLET PERFECTION** (New). Vermilion scarlet, extra large. 20/- per 100, 2/9 doz.

**SHAKESPEARE.** Rich glistening red maroon, dark centre. 15/- per 100, 2/3 per doz.

DARWIN TULIPS—SALMON KING.

**AFTERGLOW** (new). A grand new Darwin, large globular flower of a deep salmon-rose, outside petals salmon orange with amber margin. 1/3 each, 13/6 per doz.

**ANDROMAQUE.** Garnet-red, unique colour, large flower. 15/- per 100, 2/3 per doz.

**BARON DE LA TONNAYE.** Bright rose shaded blush, fine flower, lasting a long time. 14/- per 100, 2/- per doz.

**BEAUTY.** Soft rosy heliotrope with white centre. 15/- per 100, 2/3 per doz.

**CENTENAIRE** (New). Beautiful deep rose, splendid. 28/- per 100, 3/9 per doz.

**CLARA BUTT.** Large beautiful flowers, colour a lovely soft salmon rose. The Queen of Darwins. 10/- per 100, 1/6 per doz.

**DOROTHY.** Pale heliotrope, shaded white externally. 16/6 per 100, 2/4 per doz.

**DREAM.** Extra fine large flower of a beautiful shade of lilac. 15/- per 100, 2/3 per doz.

**ELECTRA.** Soft rosy lilac, with wide margin of greyish white. The most beautiful of the very pale varieties. 12/6 100, 1/6 doz.

**ERGUSTE.** Deep heliotrope, flushed silver, tinted violet inside; long beautiful flower. 14/- per 100, 2/- per doz.

**EUROPE.** Glowing salmon-scarlet, shaded rose, snow-white centre. 12/6 per 100, 1/10 doz.

**FAUST.** Rich plum-purple, very deep colour. 22/6 per 100, 3/3 per doz.

**FIRE KING.** Bright red, flower of good shape and lasting. 15/- per 100, 2/3 per doz.

**FLAMBEAU.** Brilliant scarlet with blue centre, strong grower. 14/- per 100, 2/- per doz.

**FLAMINGO.** Silvery rose-pink, very fine and distinct. 21/6 per 100, 3/- per doz.

**GALATEA.** Deep cherry carmine, brilliant effect. 20/- per 100, 2/9 per doz.

**GLOW.** Dazzling vermilion scarlet, white base with blue markings. A very large, beautiful Tulip, of wonderful brilliance of colour. 14/- per 100, 2/- per doz.

**HARRY VEITCH.** Bright brownish blood red, splendid bedder, very effective. 15/- per 100, 2/3 per doz.

**HENNER.** Brilliant chestnut brown, blue-black centre, edged white. 14/- per 100, 2/- per doz.

# DARWIN TULIPS—May Flowering.

**THE REV. EWBANK.** Silvery heliotrope shading to dove grey. 12/- per 100, 1/9 per doz.
**VAN POORTVLIET.** Brilliant carmine cerise, blue centre. 12/- per 100, 1/9 per doz.
**VIKING.** Intense rich violet, shaded purple. 18/6 per 100, 2/8 per doz.
**WHISTLER.** Vivid crimson scarlet, extra fine large flower. 20/- per 100, 2/9 per doz.
**WHITE QUEEN.** Soft rosy lilac, passing to pure white. 12/- per 100, 1/9 per doz.

## COLLECTIONS

### for Border Planting. (Our Selection).

| | |
|---|---|
| 100 in 10 beautiful varieties | 14/- |
| 50 in 10 choice sorts | 7/6 |
| 36 in 6 splendid varieties | 6/- |
| 24 in 6 fine varieties | 4/- |

## DARWIN TULIPS,

very choice mixed, in beautiful variety
10/6 per 100, 1/6 per doz.

From **Mr. H. PARSON.**
September 30th.
"The Tulips I had from you two seasons ago have been a great success, and the bulbs are still good. Everyone remarked how lovely they were."

From **Miss SCALE,** Rhiwfilin.
April 26th.
"I was very satisfied with the Bulbs, particularly the Hyacinths, the blooms were very beautiful."

DARWIN TULIPS—CLARA BUTT.

# REMBRANDT TULIPS—May Flowering.

A grand class of Tulips with large, fleshy petalled flowers, cup shaped, on long stout stems. These originated from the Darwin Tulip, which they resemble, except in colour, the flowers having broken in a permanent variegation, and including a great variety of flamed and pencilled blooms in many charming colours.

**APOLLO.** Rosy lilac and white with deep carmine stripes. 17/- per 100, 2/6 per doz.
**BOUGAINVILLE.** Amaranth and violet on white, effective. 17/- per 100, 2,6 per doz.
**CRIMSON BEAUTY.** Carmine and white, with dark cherry-red flames ; very prettily marked. 17/- per 100, 2/6 per doz.
**HEBE.** White and pale lilac, maroon markings, showy. 16/6 per 100, 2/4 per doz.
**MARCO SPADO.** White, brilliantly feathered red. 17/6 per 100, 2/6 per doz.
**QUASIMODO.** A striking and distinct variety. Deep ruby red ground, heavily flamed white, very handsome. 16/6 per 100, 2/4 per doz.
**SEMELE.** Old rose feathering on white ground, large flower. 17/- per 100, 2/6 per doz.
**SPLENDID MIXED.** All Colours. 14/- per 100, 2/- per doz.

From **Miss GODFREY,**
September 16th.
"The Tulips I had from you last year were very handsome and satisfactory."

# COTTAGE TULIPS—May Flowering.

Producing a charming display of bloom after all the other Spring bulbs have passed over, these bridge the period between the Spring and Summer flowering plants. Many of the most beautiful sorts have been recovered from old-fashioned cottage gardens, where they have been cultivated for generations. The bulbs may be planted in any good garden soil, and placed at a depth of four or five inches will thrive splendidly.

**AVIS KENNICOTT.** Rich chrome yellow, large flower. 20/- per 100, 2/9 per doz.

**BOUTON D'OR, OR GOLDEN BEAUTY.** Rich golden yellow, thick broad petals, cup-shaped flower. 12/- per 100, 1/9 per doz.

**CALEDONIA.** Brilliant orange scarlet, one of the most effective bedders, being rather dwarfer in habit than most Cottage Tulips. 11/- per 100, 1/8 per doz.

**CARNATION.** The true old English form of Picotee, but larger flowers than that variety, beautiful white heavily edged with carmine pink. 15/- per 100, 2/3 per doz.

**CYCLOPS.** Glowing scarlet, yellow base, black anthers; a lovely shaped flower. 16/6 per 100, 2/4 per doz.

**DIDIERA ALBA.** Silvery white flowers, deliciously scented. 12/6 per 100, 1/10 doz.

**DORIS.** Large globular flowers, soft rosy lilac, flushed silvery grey; a refined tulip of graceful habit. 12/- per 100, 1/9 per doz.

**ELEGANS LUTEA MAXIMA.** Magnificent rich orange-yellow, very large flower with pointed petals. 17/6 per 100, 2/6 per doz.

**ELLEN WILLMOTT.** One of the very finest. Very large beautifully shaped flowers of a soft creamy yellow colour. Deliciously fragrant. 14/- per 100, 2/- per doz.

**FAIRY QUEEN.** Deep rosy heliotrope, broadly margined amber. 12/- per 100, 1/9 per doz.

**FULGENS.** Deep crimson lake, reflexed petals. 11/- per 100, 1/8 per doz.

**GALA BEAUTY.** Exceptionally showy variety, long pointed petals; bright yellow striped scarlet. 21/- per 100, 3/- per doz.

**GESNERIANA LUTEA.** Colour rich yellow, large egg-shaped flower, tall and late blooming. 12/6 per 100, 1/10 per doz.

**GESNERIANA MAJOR.** Large rich crimson-scarlet with deep indigo blue base, splendidly effective variety. 12/6 per 100, 1/10 per doz.

**GOLDEN CROWN.** Deep golden yellow, edged and flushed crimson, very fine. 9/- per 100, 1/4 per doz.

**GOLD FLAKE.** Orange scarlet, flaked gold. 12/- per 100, 1/9 per doz.

**INGLESCOMBE PINK.** Large globular flowers, delicate rose pink with salmon glow. 14/- per 100, 2/- per doz.

**,, SCARLET.** Intense vermilion scarlet, showy handsome flower. 14/- per 100, 2/- per doz.

**,, YELLOW.** Large perfectly formed flowers of a glossy canary yellow, often described as a yellow Darwin. 12/- per 100, 1/9 doz.

**ISABELLA** (Shandon Bells). Carmine rose with creamy white shading, long flower on stem of medium height. 12/6 per 100, 1/10 per doz.

**JOHN RUSKIN.** Apricot rose, mauve shaded, egg-shaped flower, pretty. 10/- per 100, 1/6 per doz.

**LA MERVEILLE.** Deliciously scented flowers of a beautiful shade of scarlet, with apricot or orange shading. 12/- per 100, 1/9 per doz.

**LEGHORN BONNET.** Lovely pale yellow, most beautiful. 14/- per 100, 2/- per doz.

**MACROSPILA.** Glowing crimson-scarlet, black and yellow centre; sweet scented. 9/- per 100, 1/4 per doz.

**MOONLIGHT.** A long oval-shaped flower of good substance and medium height. The colour is a pleasing shade of soft yellow; fine for cutting. 15/- per 100, 2/3 per doz.

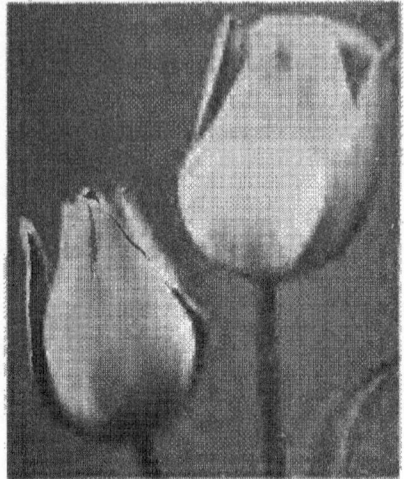

COTTAGE TULIP—INGLESCOMBE PINK.

**MRS. KERRELL** (New). Orange-pink, with amber shading, very fine. 9d. each, 7/6 doz.

**MRS. MOON.** A magnificent variety with very large flowers of a bright yellow, with reflexed outer petals. 17/6 per 100, 2/6 per doz.

**ORANGE GLOBE.** Bright orange-red with yellow shading, extra large flower; sweet scented. 15/- per 100, 2/3 per doz.

**ORANGE KING.** Large sweet-scented flowers, carried on a strong stem, and lasting a long time. The colour is a glowing orange with scarlet shading. 12/6 per 100, 1/10 per doz.

**PARISIAN WHITE.** Fine white, becoming slightly edged pink with age; capital bedder. 15/- per 100, 2/3 per doz.

**PARISIAN YELLOW.** Large bright yellow; stout petalled pointed flower, very attractive. 10/- per 100, 1/6 per doz.

**PICOTEE** (*Maiden's Blush*). Elegant pure white flowers, slightly edged carmine rose. 12/6 per 100, 1/10 per doz.

**PRIMROSE BEAUTY.** Beautifully shaped flower, delicate primrose passing off to cream; fragrant. 11/- per 100, 1/8 per doz.

**QUAINTNESS.** Long handsome flowers, mahogany on gold flushed rose. 10d. each, 8/6 per doz.

**ROSE BEAUTY.** Large flower, broad petals, rich brilliant rose. 12/- per 100, 1/9 per doz.

**ROSE POMPON.** Creamy white, flaked pink and yellow, resembling a flaked carnation in colour, semi-double. 12/6 per 100, 1/10 doz.

**ROYAL WHITE.** Large creamy white flowers with yellow centre. 15/- per 100, 2/3 per doz.

**SCARLET EMPEROR.** A fine variety with extra large bold flowers of a brilliant glowing scarlet with clear yellow base. 18/6 per 100, 2/8 per doz.

# COTTAGE TULIPS—May Flowering.

**STRIPED BEAUTY.** Rose, heavily flaked crimson and white, very showy. 15/- per 100, 2/3 per doz.

**SUNSET.** Brilliant crimson, flushed with gold ; superb variety. 14/- per 100, 2/- per doz.

**THE FAWN.** A refined and charming colour, beautiful rosy fawn, shaded blush, long egg-shaped flower. 11/- per 100, 1/8 per doz.

## COLLECTIONS for BORDER PLANTING
(OUR SELECTION).

**100 in 10 beautiful varieties, 12/6**

**50 in 10 choice sorts, 6/6**

**36 in 6 splendid varieties, 5/6**

**24 in 6 fine varieties, 4/-**

**COTTAGE TULIPS—Splendid mixed, 9/- per 100 ; 1/4 per doz.**

This fine mixture is highly recommended by us, it includes a great variety of colours made up from the best sorts.

# LATE FLORISTS' TULIPS—May Flowering.
### (Old Dutch Breeder Tulips, Bizarres, etc.)

These grand old favourites have received a good deal of attention of recent years. The combinations of colours are unusual and artistic, embracing a great range of shades in purple, bronze, lilac, heliotrope, violet, yellow, maroon, brown, gold, and orange. The flowers are mostly large, of perfect cup shape, and are carried on long, strong, upright stems. The whole class is most valuable for border work or for cut bloom, their lasting qualities making the display of great value when other flowers are scarcest.

**APRICOT.** Bronze, shaded ruddy apricot, large flower. 17/6 per 100, 2/6 per doz.

**ASPASIA.** Claret or wine-red, shaded purple. 14/- per 100, 2/- per doz.

**BACCHUS.** Rich plum-purple, very large bloom. 1/- each, 10/6 per doz.

**BRONZE QUEEN.** Enormous flower, bronze tinged apricot. 12/6 per 100, 1/10 per doz.

**DOM PEDRO.** Coffee-brown, shaded maroon, large bold flowers associate well with the lighter coloured varieties. 27/6 100, 3/9 doz.

**FEU ARDENT.** Large bold flower, Brilliant mahogany red. 16/6 per 100, 2/4 per doz.

**GENERAL NEY.** Old gold with olive base. 18/6 per 100, 2/8 per doz.

**GRAND MAITRE.** Deep purple violet, good bedder. 18/6 per 100, 2/8 per doz.

**GOLDEN BRONZE.** A most beautiful variety with large cup-shaped flowers of a rich golden bronze, shaded old gold at the base. 18/6 per 100, 2/8 per doz.

**HAMLET.** Coffee-brown with bronze edge, shaded gold. 18/6 per 100, 2/8 per doz.

**KINGSCOURT.** An enormous flower of a peculiar blend of mauve and rose, flushed a brownish tint, and edging off to almost pure orange. 12/6 per 100, 1/10 per doz.

**LA CIRCASSIENNE.** Long flower, dark old rose, feathered white, very attractive. 32/- per 100, 4/6 per doz.

**LA SINGULIERE.** Silvery white, margined crimson maroon, deepening with age until the whole flower is suffused with colour. 16/6 per 100, 2/4 per doz.

**LOUIS XIV.** Dark purple, flushed golden bronze, the finest of the Breeders. 7d. each, 6/- per doz.

**LUCIFER.** A grand tulip. Flowers terra-cotta orange, well shaped on a graceful stem. 6/6 per doz.

**MARIE LOUISE.** Old rose, shaded salmon and edged orange-bronze. 14/- per 100, 2/- doz.

**MARY HOLLIS.** Soft carmine cerise, medium size flower. 14/- per 100, 2/- per doz.

**MEDEA.** Huge flower, salmon-carmine, shaded heliotrope. 12/- per 100, 1/9 per doz.

**PANORAMA.** Deep bronzy orange red with mahogany shading. Enormous globular flowers, lasting a long time in good condition. 24/- per 100, 3/4 per doz.

**PRINCE OF ORANGE.** Globular flower, terra-cotta orange, with broad margin of a lighter shade. Showy and sweet-scented. 18/6 per 100, 2/8 per doz.

**SANSPAREILLE.** Glowing purple-violet, white base. 21/- per 100, 3/- per doz.

**SUNRISE.** An attractive colour. Reddish apricot with golden edge, dark base. 12/- per 100, 1/9 per doz.

**TERRA COTTA.** Brownish red, with inside of petals golden-orange. 14/- per 100, 2/- per doz.

**TURENNE.** Grand flower, bronzy violet purple, broad amber margin. A tall stately tulip. 17/6 per 100, 2/6 per doz.

**VIOLET QUEEN.** Beautiful reddish-violet, distinct and fine. 20/- per 100, 2/9 per doz.

**YELLOW PERFECTION.** Light bronzy yellow edged with gold. 16/6 per 100, 2/4 per doz.

**MIXTURE OF LATE FLORISTS' TULIPS.** A grand variety of colours including both flaked and striped varieties and self colours. 14/- per 100, 2/- per doz.

# NARCISSI OR DAFFODILS.

## DANIELS' "SPECIAL" 10/6 COLLECTION.

*Containing 18 splendid varieties for Exhibition or Garden decoration as below. Carriage Free, 10/6.*

| | |
|---|---|
| 3 Barbara Holmes | 3 King Edward VII. |
| 3 Cassandra | 3 Lucifer |
| 3 Duchess of Westminster | 3 Lulworth Beauty |
| 3 Emperor | 3 Madame de Graaff |
| 3 Empress | 3 Mrs. Walter Ware |
| 3 Evangeline | 3 Salmonetta |
| 3 Firebrand | 3 Seagull |
| 3 Golden Spur | 3 Sir Watkin |
| 8 King Alfred | 3 White Lady |

## DANIELS' GENERAL COLLECTIONS of NARCISSI.

### Our own Selection of Varieties.

| | | |
|---|---|---|
| 100 in 10 fine named sorts .. .. | 17s. | 6d |
| 100 in 10 popular varieties .. .. | 12s. | 6d. |
| 12 in 12 choice named sorts .. .. | 5s. | 6d. |
| 12 in 12 good named varieties.. . | 3s. | 0d. |

NARCISSUS—EMPEROR.

## TRUMPET-FLOWERED VARIETIES.

The flowers of this section have the trumpet as long, or rather longer than the divisions of the perianth.

**DUKE OF BEDFORD.** Perianth pure white, soft yellow trumpet, fine exhibition variety. 1/9 each, 20/- per doz.

**EMPEROR.** Immense size and great substance. Clear golden yellow, excellent for pots and forces well. 14/- per 100, 2/- per doz. Selected Bulbs. 20/- per 100, 2/9 per doz.

**EMPRESS.** One of the finest of the Bicolor section, perianth pure white and of great substance, trumpet rich yellow. 16/6 per 100, 2/4 per doz. Selected Bulbs. 20/- per 100, 2/9 per doz.

**GRANDEE.** Pure white perianth, rich yellow trumpet, later than Empress. 17/6 per 100, 2/6 per doz.

**GOLDEN SPUR.** Very fine, immense golden trumpet, splendid for forcing; this variety is the most used for this purpose. 15/- per 100, 2/3 per doz. Selected Bulbs. 20/- per 100, 2/9 per doz.

**GLORY OF LEYDEN.** A grand flower, with immense golden yellow trumpet, pale yellow perianth, an erect growing and stiff stemmed variety. 21/- per 100, 3/- per doz.

**KING ALFRED.** A remarkably handsome and splendid variety. The flowers which are of large size are of the most refined form and great substance; colour a clear rich golden yellow. F.C.C., R.H.S. 8d. each, 6/6 doz.

**LORD ROBERTS.** Golden yellow, broad massive perianth and enormous trumpet; very fine. 1/3 each, 13/6 per doz.

**LOVELINESS.** A flower of remarkable beauty, pure white, with expanded trumpet. 10d. each, 9/- per doz.

**MADAME DE GRAAFF.** A most superb and beautiful variety, perianth pure white, the trumpet pale primrose passing to pure white; flowers of exquisite beauty. 15/- per 100, 2/3 per doz. Selected Bulbs. 21/- per 100, 3'- per doz.

**MADAME PLEMP.** Pure white perianth, large well-formed golden yellow trumpet. 18/6 per 100, 2/8 per doz.

**MONARCH.** One of the most refined of the giant Trumpets. Beautiful golden yellow. 1/3 each. 12/6 per doz.

**MRS. J. B. M. CAMM.** White perianth, trumpet sulphur white, a very fine flower. 20/- per 100, 2/9 per doz.

**MRS. WALTER WARE.** Broad white perianth, golden yellow trumpet, a fine bicolor. 17/- per 100, 2/6 per doz.

**OLYMPIA.** Fine large yellow trumpet with paler perianth ; much larger than Emperor. Robust constitution. 10d. each, 9/- per doz.

**PRINCEPS MAXIMUS.** Immense pale yellow trumpet with sulphury divisions ; excellent for forcing. 11/- per 100, 1/8 per doz. Selected Bulbs. 15/- per 100, 2/3 per doz.

**VAN WAVEREN'S GIANT.** A gigantic variety, similar in form and colour to "Emperor," but flowers nearly three times as large. Splendid exhibition variety. 10d. each, 9/- per doz.

**VICTORIA.** Large, bold, erect flowers, perianth creamy white. Trumpet clear rich yellow. A fine variety. 14/- per 100, 2/- per doz. Selected Bulbs. 20/- per 100, 2/9 per doz.

**WEARDALE PERFECTION.** Very large flowers, perianth white, trumpet pale yellow ; magnificent variety. 7d. each, 6/- per doz., 45/- per 100.

## BARRI VARIETIES.

In this section the cup or crown measures one-third or less of the length of the perianth segments.

The beauty of this class is more fully appreciated when used as cut flowers. By cutting as they are just opening, the full beauty of their brilliantly coloured cups is best preserved.

**ALBATROSS.** Large flower, white perianth, citron yellow cup conspicuously edged orange red. 16/6 per 100, 2/4 per doz.

**BARBARA HOLMES.** White perianth, yellow cup edged bright orange-scarlet. 18/6 per 100, 2/8 per doz.

**BLOOD ORANGE.** Broad lemon coloured perianth with brilliant orange scarlet eye. 18/6 100, 2/0 per doz.

# NARCISSI OR DAFFODILS.

### Barri Varieties—*(continued).*

**BRILLIANCY.** A giant form of Conspicuus, deep yellow perianth, large cup edged deep scarlet. 3/3 each, 36/- per doz.

**CŒUR DE LION.** Brilliantly coloured and handsome flower. Cup fiery orange scarlet, perianth delicate primrose. 8d. each, 6/6 per doz.

**CONSPICUUS.** Deep primrose with orange yellow cup; very useful for cutting. 11/- per 100, 1/8 per doz.
Selected Bulbs. 14 - per 100, 2/- per doz.

**FAIR MAIDEN.** White broad perianth, crown yellow with orange-buff edge. 17/6 per 100, 2/6 per doz.

**FIREBRAND.** Creamy white perianth, shaded lemon, fluted cup of an intense fiery red. A charming decorative plant. 4d. each, 3/3 per doz., 22/6 per 100.

**GLITTER.** Orange-scarlet cup, broad yellow perianth, showy. 21/- per 100, 3/- per doz.

**MASTERPIECE.** Rounded creamy white perianth, brilliant red flattened crown. 6/- ea.

**RED BEACON.** Ivory white perianth of great substance, fiery orange-red cup. 1 6 each, 15/- per doz.

**SEAGULL.** Charming flower, large white perianth, canary cup edged apricot. 18/6 per 100, 2/8 per doz.

**STONECHAT.** A dainty Burbidgel, clear yellow edged deep orange. 22 6 per 100, 3/3 per doz.

## INCOMPARABILIS VARIETIES.

This magnificent class rivals the Barri Group in brilliancy of colouring, the trumpet is longer than that type, varying from one-third to nearly the length of the perianth. The flowers are large and all of great decorative value.

**AUTOCRAT.** Broad yellow perianth, expanded yellow cup. Free grower and showy. 15/- per 100, 2/3 per doz.

NARCISSUS—MADAME DE GRAAFF.

**Incomparabilis Varieties** *—(continued.)*

**BEDOUIN.** Large flower, white perianth, fiery orange-scarlet cup, elegantly fluted. 2/3 each, 25/- per doz.

**BERNARDINO.** Very large creamy perianth, large pale cup, heavily stained orange apricot. 3/6 each, 38/6 per doz.

**GLORIA MUNDI.** Rich yellow perianth, and large crown heavily edged orange scarlet. 17/6 per 100, 2/6 per doz.

**GREAT WARLEY.** The largest variety of Incomparabilis; white perianth, clear yellow crown, often one and a half inches across. 3/6 each, 38/6 per doz.

**HOMESPUN.** Distinct clear yellow, the most perfectly proportioned of all the Narcissus. 8d. each, 6/6 per doz.

**LUCIFER.** Large handsome white perianth, cup intense glowing orange red. 18/6 per 100. 2/4 per doz.

**LULWORTH BEAUTY.** Pure white perianth, bright orange-red cup. 15/- per 100, 2/3 doz.

**MACEBEARER.** Large creamy yellow perianth, petals overlapping; large cup pale yellow, orange red band. 5/- each, 55/- doz.

**SIR WATKIN.** Large yellow cup, tinged with orange, sulphury perianth. 15/- per 100, 2/3 per doz.
Selected Bulbs. 20/- per 100, 2/9 per doz.

**STELLA SUPERBA.** A grand bold flower, large white perianth, yellow cup. 16/6 per 100, 2/4 per doz.

**TORCH.** Yellow perianth, large bright red cup, good decorative variety. 24/- per 100, 3/4 per doz.

**WILL SCARLETT.** A fine variety. The largest and most brilliant scarlet cup of any variety; a splendid variety for exhibition. 8d. each, 7/- per doz.

From **Mrs. J. OTTY,** Weybourne.

May 8th.

" The **Hyacinths** we had from you were simply lovely, admired by all who saw them. We had nine perfect blooms from one bulb."

INCOMPARABILIS—LUCIFER.

# NARCISSI OR DAFFODILS.

## LEEDSI VARIETIES.

Known as the Euchari-flowered or Star Narcissi; this class excels by its graceful and dainty habit of bloom. The colours are white or silvery perianths, the crowns varying from white, ivory, lemon and apricot.

**BEATRICE.** Perianth and cup pure white, of fine globular form, very graceful. 15/- per 100, 2/3 per doz.

**DUCHESS OF WESTMINSTER.** Large white petals; long canary crown tinged with orange. 20/- per 100, 2/9 per doz.

**EMPIRE.** Very broad white perianth, large crown, lemon shading to white, beautifully frilled. 10/6 each.

**EVANGELINE.** Flower of fine form, broad white perianth, citron yellow cup, free bloomer. 18/6 per 100, 2/8 per doz.

**KATHERINE SPURRELL.** Long, broad, pure white perianth, of stout substance; clear lemon cup. 20/- per 100, 2/9 per doz.

**LORD KITCHENER.** Handsome Giant Leedsi. Very large flower, flat pure white perianth, cup delicate pale primrose, well opened. 2/3 each, 25/- per doz.

**MERMAID.** White perianth, widely expanded crown, opening primrose and passing off to white. 1/- each, 10/6 per doz.

**MINNIE HUME.** Broad white perianth with large spreading lemon cup. 10/- 100, 1/6 doz.

**MRS. LANGTRY.** Broad white perianth, large white crown, distinctly edged with golden yellow. 10/- per 100, 1/6 per doz.

**SALMONETTA.** A distinct and beautiful flower. Perianth clear white, cup apricot yellow passing to peach. 16/6 per 100, 2/4 per doz.

**SOUTHERN GEM.** Giant Leedsi. White with large lemon-coloured crown. 7d. each, 6/- per doz.

**WHITE LADY.** An exquisite flower with broad white perianth of perfect form, dainty cup of beautiful pale canary yellow; prettily crinkled. 4d. each, 3/- per doz., 21/- per 100.

**WHITE QUEEN.** The White "Sir Watkin." Broad pure white perianth, very large citron cup, passing to white. 7d. each, 6/- doz.

## DOUBLE-FLOWERED VARIETIES.

This old-fashioned class give a very lasting effect, the large full flowers of varying shades of yellow showing up brightly when grown in clumps on herbaceous borders or any place where they can be placed permanently.

**ARGENT.** Beautiful star shaped semi-double flowers; creamy white with yellow centre. 20/- per 100, 2/9 per doz.

**BUTTER AND EGGS** (Incomp. fl. pl.). Very large round double flowers, rich light yellow with orange yellow centre, very handsome and showy. 14/- per 100, 2/- per doz.

**ORANGE PHŒNIX** (Eggs and Bacon). Beautiful large double white flowers, with rich orange red segments. 17/- per 100, 2/6 doz.

**POETICUS PLENUS** (The Double White Gardenia-flowered Narcissus). Beautiful full double pure white flowers, deliciously scented; the last of the Narcissus to flower. This variety will bloom splendidly if planted in any cool fairly shady spot where it can remain undisturbed. 7/6 per 100, 1/3 doz.

**SULPHUR CROWN** (Codlins and Cream). Large double sulphury white flowers; very handsome; a splendid variety for cutting. 18/6 per 100, 2/8 per doz.

**TELAMONIUS PLENUS** (Double Daffodil). Magnificent double golden yellow; a fine variety for massing. 14/- per 100, 2/- per doz. Selected Bulbs, 18/6 per 100, 2/8 per doz.

## POETICUS VARIETIES.

The Poet's Narcissi all have pure white perianths, and a small flat eye or crown, lemon or yellow, edged with scarlet or crimson; in some instances the crown is almost all scarlet. All are sweet scented, the old Pheasant's Eye being the most fragrant. A deep rich soil, which does not suffer from drought, is the best for this class; naturalised in grass they produce a charming effect.

**CASSANDRA.** A gigantic Poeticus of strong growth. Broad white perianth rimmed dark red. 7d. each, 6/- per doz.

**EPIC.** Flowers three inches across, snow-white perianth, canary eye edged crimson. 7d. each, 6/- per doz.

**GLORY OF LISSE.** The finest variety of Ornatus, extra large flower, very early, pure white perianth, deep edged crimson crown. 20/- per 100, 2/9 per doz.

**HORACE.** Rounded white perianth and large eye, almost all deep scarlet. Very fine. 9d. each, 8/- per doz.

**KESTREL.** A large well rounded perianth, pure white, eye of intense crimson. 4/6 each.

**KING EDWARD VII.** Petals snow-white, beautifully shaped cup, canary yellow, edged with red; splendid variety. 20/- per 100, 2/9 per doz.

**ORNATUS.** This is the most largely grown of all the Pheasant Eye type; it is the finest of all for early flowering in pots, and when gently forced will provide flowers when they are especially valuable. Very large pure white perianth, crown edged crimson. 7/6 per 100, 1/2 per doz. Selected Bulbs, 10/6 per 100, 1/6 per doz.

**PRÆCOX GRANDIFLORA.** Very early-flowering form of Ornatus. 12/- per 100, 1/9 per doz.

**RECURVUS** (The old Pheasant-Eye Narcissus). Pure white, the crown margined with red; splendid late-flowering variety. 6/- per 100, 1/- per doz.

**SOCRATES.** The deepest edged Poeticus, solid broad white perianth, cup deeply edged madder-scarlet. 3/6 each, 38/6 per doz.

# NARCISSI OR DAFFODILS.

## POETAZ NARCISSUS.

**ADMIRATION.** A new handsome variety; yellow perianth, cup heavily margined scarlet. 24/- per 100, 3/4 per doz.

**ELVIRA (A.M., R.H.S.).** Broad white petals of great substance, golden yellow cup, edged with orange, delicate fragrance. Undoubtedly the finest of all for general culture, and one we highly recommend. The finest Narcissus for howl culture. 16/6 per 100, 2/4 per doz.

**IDEAL.** White, with dark orange eye, large truss; most striking colour. 17/- per 100, 2/6 per doz.

**LOUISE.** Pure white, yellow cup, large flower, 16/6 per 100, 2/4 per doz.

**MRS. ASQUITH.** White perianth, deep yellow cup. 18/6 per 100, 2/8 per doz.

**TRIUMPH.** Deep yellow eye, perianth pure white; sterling variety, one of the best. 21/- per 100, 3/- per doz.

## POLYANTHUS VARIETIES.

A beautiful free-flowering class of easy cultivation, deliciously scented, and admirably suited for growing in pots. They produce handsome trusses of elegantly-formed flowers varying in colour from deep orange and primrose to the purest white. With the exception of the Double Roman and Paper White varieties, they will succeed well planted out of doors. For pot cultivation treat as recommended for Hyacinths.

**DOUBLE ROMAN.** Double, white, with orange yellow nectary; excellent variety for early forcing. 16/6 per 100, 2/4 per doz.

**GRAND MONARQUE.** White, lemon cup, large trusses of bloom 17/6 per 100, 2/6 doz,

**GRAND PRIMO.** White with citron cup. 15/- per 100, 2 3 per doz.

**GRAND SOLEIL D'OR.** Yellow, orange cup; fine distinct variety; one of the best varieties. 18/6 per 100, 2/8 per doz.

**PAPER WHITE, NEW LARGE-FLOWERED.** Splendid variety, its large trusses of pure white flowers may be had in abundance at Christmas or early in the New Year. This variety is the most generally used of all the Polyanthus varieties. 21/- per 100, 3/- doz.

**WHITE PEARL.** Large trusses of flowers, pure white, cup slightly shaded citron when opening. 15/- per 100, 2/3 per doz.

## JONQUILS.

Deliciously fragrant varieties of a graceful habit of growth, much esteemed for pot culture, but are quite hardy, and may be planted in the open border; very useful for cutting.

**DOUBLE CAMPERNELLI MAJOR.** Beautiful double, deep golden yellow flowers, two and three on a stem, very free bloomer. 10/- per 100, 1/6 per doz,

**SINGLE SWEET-SCENTED.** Light elegant growth, bearing charming clusters of small rich yellow flowers, flowering outdoors in May; of pleasing effect when grown in pots or bowls; may be forced very early into bloom. 4/6 per 100, 9d. per doz.

**SINGLE CAMPERNELLI MAJOR** Bright golden yellow; finely scented, first-class for cutting; elegant rush-like foliage. Much stronger growing than the ordinary single Jonquil. 9/- per 100, 1/4 per doz.

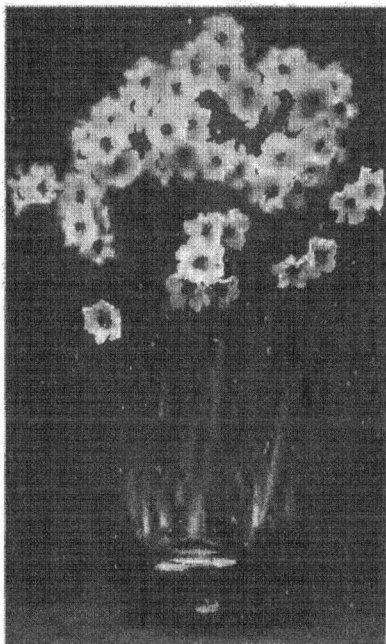

POETAZ NARCISSUS—ELVIRA.

## MINIATURE NARCISSUS.

These charming little Narcissus are most suitable for the now highly popular Rock Gardening, and are also beautiful when grown in pots.

**BULBOCODIUM CONSPICUUM.** Rich golden yellow. 9/- per 100, 1/4 per doz.
  **„ CITRINUM.** Pale yellow. 11/- 100, 1/8 doz.
  **„ MONOPHYLLUM.** Pure white. 42/- per 100, 5/6 per doz.

**CERNUUS PULCHER.** Silvery white perianth and trumpet, both of the same length, early flowering, best in a partially shaded position. 20/- per 100, 2/9 per doz.

**CYCLAMINEUS.** Lemon, reflexed petals. 21/- per 100, 3/- per doz.

**JOHNSTONI, QUEEN OF SPAIN.** Soft clear yellow, of drooping habit, with reflexed perianth and long trumpet. 25/- 100, 3/6 doz.

**MINIMUS.** Smallest Trumpet Narcissus, rich yellow, height three inches. 35/- per 100, 3/8 per doz.

**MOSCHATUS OF HAWORTH.** The snow-white Spanish Daffodil; charming; grows best in a partially shaded position. 20/- per 100, 2/9 per doz.

**TRIANDUS ALBUS.** White perianth and cup, very pretty. 10/- per 100, 1/6 per doz.

## DANIELS' CHOICE MIXTURES OF NARCISSI.

**DANIELS' CHOICE MIXED SINGLE** Trumpet-flowered varieties only; a grand mixture, including many of the very finest sorts; splendid for planting in Woodlands or Herbaceous borders, 12/6 per 100, 1/10 per doz.

**DANIELS' CHOICE MIXTURE OF ALL VARIETIES,** containing Incomparabilis, Leedsi, and Poeticus Varieties, including the most showy and popular, 9/- per 100, 1/4 per doz.

# CROCUS—Splendid Named Varieties.

Much skill has been devoted during recent years to the hybridisation of Crocuses, and we are able to offer several new and charming sorts.

The Crocus is splendid for pot culture, bowls in fibre, and window boxes.

**ALBION.** Purple ; fine large flower. 4/- por 100, 9d. per doz.

**DOROTHY.** Light porcelain blue, a fine self colour. 4/- per 100, 9d. per doz.

**DISTINCTION.** Soft reddish pink, most distinct colour. 8/- per 100, 1/3 per doz.

**FANTASY.** Outside striped blue on white ground. 8/- per 100, 1/3 per doz.

**HERO.** Dark shining purple, the largest yet raised. 5/- per 100, 10d. per doz.

**KING OF THE BLUES.** Splendid large dark blue. 4/- per 100, 9d. per doz.

**LA MAJESTEUSE.** Beautiful violet striped. 5/- per 100, 10d. per doz.

**MIKADO.** Pale greyish lilac, inside of petals striped with deep mauve. 9/- per 100, 1/4 per doz.

**MINERVA.** Very finely striped blue ; one of the best. 9/- per 100, 1 4 per doz.

**NON PLUS ULTRA.** Large purple flowers, conspicuously edged with white. 5/- per 100, 10d. per doz.

**PALLAS.** Light lilac stripes on white ground, extra large flower. 8/- per 100, 1/3 por doz.

**PEARL.** Very pale lilac striped, extra large. 9/- per 100, 1/4 per doz.

**QUEEN OF SHEBA.** Rich golden yellow. 9/- per 100, 1/4 per doz.

**QUEEN OF THE WHITES.** Very large, pure white. 4/- per 100, 9d. per doz.

**SIR WALTER SCOTT.** Blue and white striped. 4/- per 100, 9d. per doz.

**WHITE GIANT.** The largest pure white, very fine. 8/- per 100, 1/3 per doz.

**SPLENDID MIXED, from above-named varieties** 4/- per 100, 9d. per doz.

CROCUS, MIKADO.

## COLLECTIONS OF CHOICE CROCUS.
### OUR OWN SELECTION.

100 in 10 choice varieties, with names, 6/-.
50 in 10 choice varieties, with names, 3/6.

## CROCUS—Showy Varieties for Massing.

Planted in broad marginal lines as edgings, or in large masses of such distinct colours as dark blue and yellow, blue and white, &c., these are strikingly effective. They are also highly recommended for any waste place where they may remain undisturbed. They will produce a beautiful display each succeeding Spring.

**BLUE.** Various shades, mixed. 2/6 per 100, 6d. per doz.

**WHITE.** Fine pure white. 2/6 per 100, 6d. per doz.

**STRIPED.** Various shades. 8/- per 100, 8d. doz.

**YELLOW.** Extra large roots ; fine. 6/- per 100, 1/- per doz.

**YELLOW.** Fine golden yellow. 4/- per 100, 9d. per doz.

**MIXED.** All colours. 2/6 per 100, 6d. per doz.

## CROCUS SPECIES.

**CROCUS, CLOTH OF GOLD (Susianus).** Yellow with brown stripe, free flowering. 6/- per 100, 1/- per doz.

" **IMPERATI.** Winter flowering, violet, fawn outside, feathered. 14/- per 100, 2/- per doz.

" **SPECIOSUS.** Autumn flowering, bright blue-violet, large and showy. 14/- per 100, 2/- per doz.

" **TOMASSINIANUS.** A beautiful Winter flowering species, blooming even before the snowdrop. The flowers are a pale lavender colour, the outside a silvery grey ; very showy. 8 - per 100, 1/3 per doz.

" **ZONATUS.** Autumn flowering, rosy lilac, orange zone. 8/- per 100, 1/3 per doz.

# DANIELS' SUPERB RANUNCULI.

For early flowering, plant the tubers (claws downwards) in November or December in drills about six inches apart, two inches deep, with about three or four inches between the tubers in the row ; cover with fine soil and rake level.

### DANIELS' GIANT TURBAN.

A vigorous and free-flowering class, each plant when well grown producing from forty to fifty large double flowers of the most brilliant colours ; useful for cutting.

**MONT BLANC.** Splendid pure white. 8/- per 100, 1/3 per doz.

**PRIMROSE BEAUTY.** Lovely shade of yellow. 6/- per 100, 1/- per doz.

**VESUVIUS.** Fine scarlet. 8/- per 100, 1/8 doz.

**MIXED.** Great variety of colour. 5/- per 100, 10d. per doz.

### DANIELS' GIANT FRENCH.

A magnificent class, producing very large double and semi-double flowers of the richest and most brilliant colours ; splendid for cut flowers and last a long time in water. These are very superior to the Persian.

**SUPERB MIXED.** 10/- per 100, 1/6 per doz.

**PERSIAN MIXED.** A fine mixture, flowers of faultless form. 5/- per 100, 10d. per doz.

**TURBAN MIXED.** Strong growing, many colours. 5/- per 100, 10d. per doz.

**TURBAN SCARLET.** Very showy for ribbon borders or beds. 7/- per 100, 1 2 per doz.

# DANIELS' CHOICE NAMED IRISES.

GROUP OF IRISES,

## ENGLISH IRIS.

The flowers of this beautiful class are larger than those of the Spanish varieties, and include a charming range of colours, many of the sorts having handsomely blotched or mottled flowers. They are very free flowering, perfectly hardy, and splendid for cutting.

**CELESTIAL BLUE.** Beautiful sky blue. 12/- per 100, 1/9 per doz.
**KOH-I-NOOR.** White, splashed and mottled rose purple, very effective. 12/6 100, 1/10 doz.
**MONT BLANC.** Creamy white. 12/- 100, 1/9 doz.
**PERLE DES JARDINS.** Lovely pearl blue with light blue markings. 12/6 100, 1/10 doz.
**PRINCE OF WALES.** Rich dark blue. 12/- per 100, 1/9 per doz.
**ROSA BONHEUR.** White, splashed carmine and purple. 12/- per 100, 1/9 per doz.
**SPLENDID MIXED.** In beautiful variety. 8/- per 100, 1/3 per doz.

## IRIS SPECIES.

**DANFORDIÆ.** The golden Reticulata, flowering in February. 4/6 per doz.
**HISTROIDES.** Light blue, varying in shade, spotted ; larger flower than Reticulata and very early. 6/- per doz.
**KÆMPFERI.** Beautiful Japanese Irises ; wonderfully decorative. Choicest Mixed, Double and Single-flowered. 3/6 per doz.
**PAVONIA** (Peacock Iris). Pure white, spotted with delicate blue. 12/- per 100, 1/9 per doz.
**REGELIO-CYCLUS, CHARON.** Golden brown, feathered chocolate. 1/6 each, 15/- per doz.
" **HECATE.** Rosy lilac with brown veinings. 2/- each, 21/- per doz.
" **HERA.** Ruby red, bronze, and blue. 2/- each. 21/- per doz.
" **OSIRIS.** Satiny white veined lilac purple. 1/6 each, 15/- per doz.
**STYLOSA.** Light blue, winter flowering. 2/- doz.
**SUSIANA.** A handsome species, greyish flower, finely veined with dark lines. 4/6 per doz.
**TUBEROSA** (Snake's Head Iris). Velvety-green and black flowers. 12/6 100, 1/10 doz.
**RETICULATA.** See page 42

## SPANISH IRIS.

These highly popular flowers bloom a fortnight earlier than the English varieties, and should be grown freely in every garden. The colours range from the richest yellow through all the shades of blue, bronze, lilac, etc., to the purest white, and the flowers last a long time when out and placed in water. All are very hardy and free flowering. Plant about three inches deep, any time from September to December.

**BEAUTY.** Pale lavender blue, white falls with orange blotch. 4/- per 100, 9d. per doz.
**BELLE CHINOISE.** Lovely deep yellow. 5/- per 100, 10d. per doz.
**BLANCHE SUPERBE.** Splendid pure white. 4/- per 100, 9d. per doz.
**BRONZE KING.** Fine golden orange, tinted purple. 6/- per 100, 1/- per doz.
**CAJANUS.** Canary yellow, orange yellow blotches. 5/- per 100, 10d. per doz.
**EXCELSIOR.** Beautiful light blue ; the latest to flower. 5/- per 100, 10d. per doz.
**FLORA.** Creamy white with pale lavender standards. 5/- per 100, 10d. per doz.
**GOLD CUP.** Splendid bronze and yellow ; the finest variety. 14/- per 100, 2/- per doz.
**HERCULES.** Very fine, bronze and violet. 5/- per 100, 10d. per doz.
**KING OF THE BLUES.** Dark blue ; very fine. 5/- per 100, 10d. per doz.
**KING OF THE WHITES.** The finest pure white, orange blotch. 6/- per 100, 1/- doz.
**LA NUIT** (New). Splendid dark blue, very distinct. 6/- per 100, 1/- per doz.
**L'UNIQUE.** Quite distinct ; violet blue, falls white blotched golden. 6/- per 100, 1/- doz.
**LA RECONNAISSANCE.** Dark bronze, golden blotch, extra fine. 5/- per 100, 10d. per doz.
**LEMON QUEEN.** Lemon yellow, tinted lavender ; early. 6/- per 100, 1/- per doz.
**PRINCE HENRY.** Very large bronze purple, the best in this shade. 5/- 100, 10d. doz.
**SOLFATERRE.** Deep purple blue, deep yellow blotch. 5/- per 100, 10d. per doz.
**SNOWBALL.** Beautiful white. 4/- per 100, 9d. per doz.
**SURBITON.** Deep rich yellow ; fine. 6/- per 100, 1/- per doz.
**WALTER T. WARE.** Creamy yellow ; fine. 6/- per 100, 1/- per doz.

Collection of 100 in 10 splendid varieties, 5/6
Collection of 50 in 10 splendid varieties, 3/-

**CHOICEST MIXED.** All colours. 2/6 per 100, 6d. per doz.

## NEW GIANT SPANISH IRIS.

A splendid and very decorative class of new hybrids far surpassing the Spanish Iris in size of bloom, blooming also about a fortnight earlier.

**ANTON MAUVE.** Beautiful pearl blue ; lovely. 9/- per 100, 1/4 per doz.
**FRANS HALS.** Pale blue standards, falls pale primrose. 12/- per 100, 1/9 per doz.
**HOBBEMA.** Light blue standard, primrose falls, large flower. 11/- per 100, 1/8 per doz.
**IMPERATOR.** Large flower, brilliant blue ; can be forced very early. 16/6 per 100, 2/4 per doz.
**REMBRANDT.** Vigorous grower, broad dark blue standards, deep blue falls with orange blotch. 12/- per 100, 1/9 per doz.
**MIXED.** A very choice mixture of seedlings. 8/- per 100, 1/3 per doz.

ANEMONE—HIS EXCELLENCY.

## DANIELS' SUPERB ANEMONES.

**ST. BRIGID.** A brilliant and very beautiful class of large semi-double flowers of the most striking and charming shades of colour, ranging from crimson and scarlet to rose, lilac, dark blue, &c., to the purest white. They are first-class for cut flowers, and if cut when the bloom is beginning to open, will retain their beauty for a long time. Very Choice Mixed. 14/- per 100, 2/- per doz.

**ANEMONE APPENINA.** The Blue Mountain Wind Flower; a charming little variety with lovely sky-blue flowers. 6/- per 100, 1/- doz.

**ANEMONE FULGENS.** The Scarlet Wind Flower. Beautiful brilliant scarlet single flowers, blooming early in the year. Prefers a rich well-drained soil. 17/- per 100, 2/6 doz.

## GIANT FRENCH.

We can thoroughly recommend our strain of Giant French Anemones, the large flowers are of the most varied and brilliant colours.

**HIS EXCELLENCY.** Brilliant deep scarlet, thick petals and strong stems. 12/- per 100, 1/9 doz.

**SINGLE MIXED.** Great variety. 10/6 per 100, 1/6 per doz.

## DOUBLE-FLOWERED DUTCH.

A beautiful class of double flowers, very useful for garden decoration.

**KING OF THE BLUES.** Lovely dark blue. 12/- per 100, 1/9 per doz.

**KING OF THE SCARLETS.** Brilliant scarlet. 12/- per 100, 1/9 per doz.

**FINEST MIXED.** Double. 10/6 per 100, 1/6 doz.

## SINGLE-FLOWERED DUTCH.

Beautiful large-flowered varieties. Valuable for cut flowers.

**FINEST MIXED.** Single. 5/- per 100, 10d. doz.

## IXIAS—(African Corn Lily).

This wonderfully brilliant class of South African bulbs gives a profusion of graceful wiry spikes whether grown as a greenhouse subject or in the open garden. If grown outdoors the bulbs should not be planted before October or November, so that the growth may not be injured by Spring frosts.

**AZUREA.** Azure-blue, dark eye, distinct. 11/- per 100, 1/8 per doz.

**BEAUTY OF NORFOLK.** Bright yellow, maroon eye. 8/- per 100, 1/2 per doz.

**BRIDESMAID.** Globular white flowers, carmine eye. 10,- per 100, 1/6 per doz.

**DESDEMONA.** Rose and purple with black eye. Vigorous grower. 9,- per 100, 1/4 per doz.

**GOLDEN DROP.** Beautiful rich yellow. 10/- per 100, 1/6 per doz.

**HECTOR.** Rich crimson claret, fine variety. 10/- per 100, 1/6 per doz.

**HOGARTH.** Cream, dark brown eye. 8/- per 100, 1/2 per doz.

**HUBERT.** Coppery red, black eye, very fine. 8/6 per 100, 1/3 per doz.

**PRESTIOS.** Pure white, bright red eye; one of the finest. 7/- per 100, 1/- per doz.

**ROSEA PLENA.** Beautiful soft rose, double flower. 10/6 per 100, 1/6 per doz.

**VIRIDIFLORA.** The Green Ixia Sea-green, black eye. 17/- per 100, 2/6 per doz.

**VOLUNTEER.** Pale yellow, dark red eye. Strong growing. 8/- per 100, 1/2 per doz.

**WHITE SWAN.** Lovely pure white with indigo blue eye. 10/6 per 100, 1/6 per doz.

**WILLIAM THE CONQUEROR.** Beautiful creamy white, brown eye. 10/6 per 100, 1/6 per doz.

50 in 5 choice varieties, our selection, 4/-.

24 in 4 beautiful varieties, our selection, 2/6.

**CHOICE MIXED.**    4/6 per 100, 9d. per doz.

IXIA—BEAUTY OF NORFOLK.

# DANIELS' CHOICE NAMED GLADIOLI.

The early-flowering section of Gladioli—blooming in June and July—will thrive in almost any soil or situation, but should for preference be planted in a warm sunny position. October and November are the best months for planting, and the bulbs should be planted in clumps or patches of six or eight, and at a depth of four or five inches, covering them over in Winter with some short manure to prevent injury from frost. For growing in pots, they should be potted five in a six-inch pot, using a light rich compost and covering the crowns about half an inch deep. Give but little water till they start into growth, when they should be placed on the stage, or as near the light as convenient.

## SELECT GLADIOLI.
### Early-Flowering Varieties.

**ACKERMANNI.** Salmon, flaked carmine, with violet eye. 14/- per 100, 2/- per doz.
**BLUSHING BRIDE.** Lovely white, with pink and carmine flakes on lower petals, beautiful. 10/- per 100, 1/6 per doz.
**BRILLIANT** (New). Extra fine deep scarlet. 12/6 per 100, 1/10 per doz.
**CRIMSON QUEEN.** Fiery orange-scarlet with crimson glow, blotched carmine and white; very showy. 14/- per 100, 2/- per doz.
**FIERY KNIGHT.** Brilliant scarlet and vermillion, a large and beautiful flower. 21/- per 100, 3/- per doz.[
**GENERAL SCOTT.** Lovely satin rose, cream-coloured blotches, edged scarlet. 14/-100, 2/- dz.
**LIBERTY** (New). Lovely pink, carmine edged flakes. 16/6 per 100, 2/4 per doz.
**PEACH BLOSSOM.** Delicate peach pink, extra fine. 10/- per 100, 1/6 per doz.
**RED PRINCE.** Orange scarlet with white blotch. 16/6 per 100, 2/4 per doz.
**SALMON QUEEN.** Beautiful salmon-red, with white and crimson flakes. 14/- per 100, 2/- per doz.
**SWEETHEART** (New). Pure white with red blotch. 17/- per 100, 2/6 per doz.
**THE BRIDE.** Pure white; a gem for outting and forces well. 16/6 per 100, 2/4 doz.
**VERY CHOICE MIXED.** A fine mixture of these delightful flowers, containing many lovely colours. 8/- per 100, 1/3 per doz.

*Bulbs of the Early-Flowering Gladioli are sent out from October onward.*

GLADIOLUS—THE BRIDE.

For the later flowering varieties of Gladiolus, see our "Guide," published in January, which contains the names of all the finest and most beautiful kinds.

## LILIUM CANDIDUM (The Madonna Lily).

This charming old favourite has been very largely grown during the past few years as a pot plant. It will bear gentle forcing extremely well, and if planted early will bloom at a time when pure white flowers, and especially Lilies, are extremely valuable ; it is perfectly hardy, and planted in a garden will give a lovely show of its graceful spikes of bloom sooner than any of the other kind of Lilies.
**EXTRA STRONG BULBS, 10/6 per doz., 1/- each.**    **ORDINARY SIZE, 7 6 per doz., 9d. each**

## MUSCARI (Grape Hyacinths).

These charming bulbs are excellent for giving a display of flowers in the borders in spring. They are quite easy of culture, and will succeed in almost any soil, forming masses of colour in the rookery, or if grown in pots or bowls in the greenhouse, are splendid for decoration.

**ATLANTICUS.** Lovely blue, rare and distinct. 6/- per 100, 1/- per doz.
**AZUREUS.** Flowers in February in the open garden. Lovely sky-blue. 6/- 100, 1/- doz.
**PLUMOSUM MONSTROSUM** (Feather Hyacinth). Large plumes of beautiful bright purple flowers. 8/- per 100, 1/2 per doz.

**HEAVENLY BLUE.** Rich blue, most effective in masses ; very fragrant, and most useful either for decoration or outting purposes. This is the finest of all the Muscari for outdoor planting, in grass, or for bowl culture. 5.- per 100, 10d. per doz.
Selected Bulbs, 7/- per 100 ; 1 2 per doz.

## LILIUM CHALCEDONICUM (Scarlet Turk's Cap Lily).

This fine old variety has now become very scarce. The flowers are of intense scarlet with reflexed petals. Bloom in July. Each 2/6, per doz, 27/6

# MISCELLANEOUS BULBS.

## ALSTRŒMERIA.

**CHILENSIS.** Showy plants for a warm dry border. Planted nine inches deep they may remain undisturbed for years, and will bloom abundantly. Choice mixed, 3/6, per doz.

## AMARYLLIS BELLADONNA.
### (Belladonna Lily).

Delicate pale rose, autumn blooming. 1/- each, 10/6 per doz.

## ARUM LILY (Æthiopicum).

Dormant Corms. Pure white semi-dwarf variety. 6d. each, 5/- per doz.

## ARUM (RICHARDIA) ELLIOTIANA.

THE YELLOW ARUM, Beautiful colour with silver spotted leaves. Strong corms, 2/- each.

## BULBOCODIUM VERNUM.

For edgings, etc. Bright rosy purple, very early. 15/- per 100, 2/3 per doz.

## CAMASSIA ESCULENTA.

Spikes of pretty star-like flowers. Colour light blue. 7/- per 100, 1/2 per doz.

## CHIONODOXA.

This is one of the most beautiful hardy Spring-flowering bulbs in cultivation, and one of the very easiest to grow. Planted six or eight in a five-inch pot has a very pretty effect in the greenhouse. Out of doors they should be placed not less than four inches deep, and about three inches apart, and to be effective, not less than ten to twelve should be planted in a patch.

**LUCILLÆ** (Glory of the Snow). Quite hardy, will thrive in any soil, and produces an abundance of brilliant sky-blue, white-centred flowers. 6/- per 100, 10d. per doz.

**SARDENSIS.** Beautiful deep blue flowers with a small white centre; charming variety. 7/- per 100, 1/1 per doz.
Selected Bulbs. 9/- per 100, 1/4 per doz.

## CROWN IMPERIALS.
### (Fritillaria Imperialis).

These stately growing hardy plants grow well in any good garden soil, and if allowed to remain undisturbed for several years, will form picturesque groups of rare beauty. The bulbs should be planted with their crowns four or five inches below the surface, and eight ins. to a foot apart.

**SINGLE RED.** Fine showy sort. 10/6 per doz., 1/- each.
**SINGLE YELLOW.** Large flowers. 10/6 per doz., 1/- each.
**MIXED.** 4/6 per doz., 6d. each.

## FRITILLARIA (Snakeshead).

**MELEAGRIS MIXED.** Handsomely chequered flowers, hardy, grows well in howls. 6/6 per 100, 1/- per doz.

## IXIOLIRION PALLASII.

Deep blue tubular flowers, blooming in May. 8/- per 100, 1/2 per doz.

## LILY OF THE VALLEY.

Clumps for outdoor planting, ready end of October. 1/9 each, 18/6 per doz.

## LEUCOJUM VERNUM.

**SPRING SNOWFLAKE.** Large white flowers tipped green, resembling a giant snowdrop. 12/- per 100, 1/9 per doz.

## PUSCHKINIA LIBANOTICA.

**LEBANON SQUILL.** White and soft blue flowers, a charming flower. 6/6 100, 1/- doz.

## SCHIZOSTYLIS COCCINEA.

Spikes of crimson flower like a small Gladiolus, blooming during Autumn and Winter. 14/- per 100, 2/- per doz.

## SPIRÆA MULTIFLORA COMPACTA.

Clumps for pots, growing freely with ample water supply. Ready in November. 10d. each, 9/- per doz.

## SCILLA SIBIRICA.

This beautiful class should be grown freely in every garden. In height they do not exceed four or five inches, and their lovely bending sprays of rich ultra-marine blue flowers appear in the greatest profusion during the month of March. Planted eight or ten in a six-inch pot and treated as recommended for Hyacinths, they succeed equally well, and have a pretty effect in the greenhouse.

**SIBIRICA (Præcox).** Fine bright blue, splendid for edgings, clumps, or the mixed border, pots, etc. 8/- per 100, 1/2 per doz.
Selected Bulbs, 11/- per 100, 1/8 per doz.

## SCILLA CAMPANULATA
### (The Wood Hyacinth).

These lovely flowers are most showy in May in the wild garden. They also do exceedingly well planted five or six in a pot for greenhouse cultivation.

**ALBA.** 7/- per 100, 1/2 per doz.
**BLUE QUEEN.** Large bright blue flowers on strong stems. 8/- per 100, 1/3 per doz.
**CÆRULEA.** Blue spikes of flowers. 6/6 per 100, 1/- per doz.
**ROSEA.** Pretty shade. 9/- 100, 1/4 doz.

## SNOWDROPS.

Snowdrops will thrive in almost any soil or situation, and best when planted in clumps and left to take care of themselves for several years in succession, when they will form handsome and increasingly large groups. October is the best month to plant, and the bulbs should be placed three or four inches deep, and about two inches apart. It is not possible to ensure the single or double snowdrops free from a slight admixture of either variety.

|  | per 100. | | per doz. | |
|---|---|---|---|---|
|  | s. | d. | s. | d. |
| FINE DOUBLE .. | 6 | 0 | 0 | 10 |
| „ Large bulbs | 8 | 0 | 1 | 2 |
| FINE SINGLE .. | 4 | 6 | 0 | 8 |
| „ Selected bulbs | 6 | 0 | 0 | 10 |

## SPARAXIS.

Showy colours resembling Ixias, but with larger flowers and more variation of colour in the individual blooms. 5/- per 100, 10d. per doz.

## TRITELIA.

**UNIFLORA.** Early flowers, lilac white, good for edgings. 4/- per 100, 8d. per doz.
**VIOLACEA.** Lavender striped violet. 5/- per 100, 10d. per doz.

## WINTER ACONITE.

The little Aconite, which blooms earlier than the Snowdrop, is invariably the first flower in the garden to greet us in the New Year, and should always be grown freely in sunny positions near the house, or walks, for the welcome display they make with their golden-yellow flowers so early in the season.
**Strong Flowering Roots, 3/6 per 100, 7d. per doz.**

# HORTICULTURAL SUNDRIES.

## SPRAYING MACHINES.

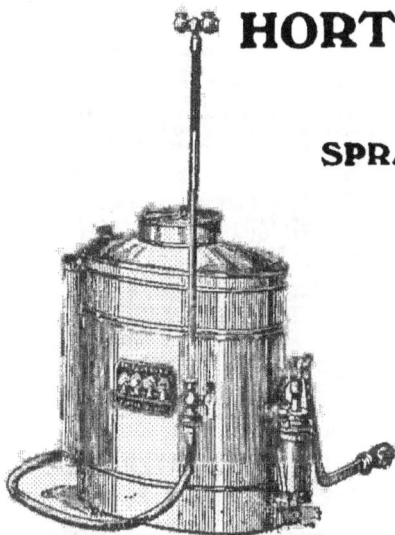

"FOUR OAKS" KNAPSACK SPRAYER.

**KNAPSACK SPRAYERS.** This machine is of convenient size. Capacity 3 gallons. Easy to use, and can be supplied either with internal or external pump. Complete with double swivel fine spraying nozzle, price £4 10s. 0d.

**PNEUMATIC SPRAYER, TYPE B.B.** Capacity 2½ gallons. Working 2 gallons. Medium size with pressure gauge 40 in. hose 20 in. spray rod. Price nett £5 3s. 0d.

**PNEUMATIC HAND SPRAYER, TYPE D.** Capacity 5 pints. Strong and light. Complete with powerful self-contained Pump, automatic valve. Price nett £2 15s. 0d.

### PORTABLE SPRAYING AND LIME WASHING MACHINES.

**THE "DUKE."** Capacity 10 gallons. Fitted with powerful pump and 10 ft. length of hose, brass spraying lance, complete with interchangeable nozzles for lime washing and fine spraying, price £11.

**BAMBOO LANCE.** 6 ft. 6 ins., for spraying tall trees, for use with the above, price 25/-.

THE **" VISCOUNT."** A very useful sprayer, holding 6 gallons. Complete with container and brass strainer, powerful pump and all accessories, price £4 12s. 6d.

SYRINGES. THE **"ABOL."** 17/6 and 22/- each. Postage 9d. extra.
 Bends for spraying under foliage, 2/- each.
 ,, **PEERLESS THREE WAY.** No bend required. 17/6 and 19/6 each. Postage 9d. extra.

## INSECTICIDES, &c.

" ABOL " INSECTICIDE. Most effectual, and may be used by any amateur. 2/2 per pint; 3/6 per quart; 5/6 per half-gallon; 10/- per gallon.
EWING'S MILDEW COMPOSITION 2/6 per bottle.
GISHURST COMPOUND. Especially useful for the winter cleansing of fruit trees. In boxes, 1/9 and 4/6 each.
KATAKILLA POWDER. To make 10 gallons Insecticide, 2/- per box ; larger size, 6/-.
MEALY BUG DESTROYER. 1/6 per jar.
McDOUGALL'S INSECTICIDE FUMERS. Sufficient for 1,000 cubic ft. 1/3 each; 15/- per doz. for 2,000 ft., 2/- each ; 24/- per doz.
TOBACCO POWDER. 9d., 1/6, and 3/6 per tin.
VAPORITE. For destroying Wireworms, &c., in soil. Tins, 1/6 and 3/9 ; 14 lbs. 5/6 ; in bags, 28 lbs., 8/- ; ¼ owt., 11/9 ; 1 owt., 18/-.
WORM DESTROYER FOR LAWNS. 1 owt., 27/6 ; ½ owt., 15/- ; 28 lbs., 8/- ; 14 lbs., 4/6.
XL ALL INSECTICIDE. The safest and most effectual Insecticide. 4/- per pint ; 7/3 per quart ; 12/9 per half-gallon ; 23/- per gallon.
XL ALL VAPORISING FUMIGATOR. In bottles sufficient for 5,000 cubic feet, 4/3 ; 10,000 7/6 ; 20,000 13/6 ; 40,000 25/-. "XL All" Fumigators for same, 2/7 for 2,000 ft. ; 3/- for 5,000 ft.

## SPECIAL WINTER WASHES, &c.

XL ALL WINTER WASH. In tins sufficient for making 8 galls., 1/6 per tin. Postage 9d. extra.
V.I. FLUID. Use 1 part to 100 parts of water. Per quart 3/9, postage 9d. extra ; ½ gall. 6/9, packing and carriage 1/6 extra ; 1 gall. 11/6, packing and carriage 2/6 extra.
BANDING GREASE. Tins to band 15 to 20 average trees 2/6 each, postage 6d. ; larger tins 8/6 each, postage 1/-.
GREASE BANDS. In packets of 20, for use with above. Per packet 6d.

## DANIELS' SUPERIOR MUSHROOM SPAWN.

Bricks, 9d. each ; 4 bricks, 2/9 ; postage on 1 brick, 9d. ; on 4 bricks, 1/3.
 Complete Instructions for Cultivation will be sent with every order.

 *Prices of all Manures and Sundries are subject to alteration without notice.*

# ARTIFICIAL MANURES, &c.

**DANIELS' NORWICH FERTILIZER.**
We have received many testimonials of the very excellent quality of this Manure, and the increasing demand is proof of its being the very best value on the market. For Chrysanthemums, Cucumbers, Grapes, Greenhouse Plants, Roses, Fruit Trees, and all General Garden Crops it is invaluable.
7 lb. 6/- ; 14 lb. 5/- ; 28 lb. 9/6 ; 56 lb. 17/- ; 1 cwt. 32/-.
**CLAY'S FERTILIZER.** In bags, 7 lb. 3/9, 14 lb. 6/6, 28 lb. 11/-, 56 lb. 20/-, 1 cwt. 36/-.
**ICHTHEMIC GUANO.** In tins, 9d. and 1/3 each. In bags, 7 lb. 3/-, 14 lb. 5/6, 28 lb. 10/-, ½ owt. 17/6, 1 owt. 62/-.
**THOMSON'S VINE, PLANT, AND VEGETABLE MANURE.** In tins, 1/6 ; 7 lb. 3/- ; 14 lb. 5/6 ; 28 lb. 9/6 ; 56 lb. 17/- ; per cwt. 32/-.
**THOMSON'S CHRYSANTHEMUM MANURE.** 7 lb. 3/-, 14 lb. 5/6, 28 lb. 9/6, 56 lb. 17/-.
**TOMORITE.** The finest Manure for Tomatoes ever manufactured. In cartons 9d. and 1/3, 7 lb. 6/-, 14 lb. 5/6.
**BASIC SLAG.** Per cwt. 6/6, ½ cwt. 5/-, 28 lb. 3 -.
**BONE MEAL.** Per cwt. 16/-, ½ cwt. 10 -, ¼ cwt. 5/6, 14 lb. 3/-.

## POTTING MATERIALS.

**CHARCOAL.** In lumps 6/- per bushel.
**LEAF SOIL.** 3/6 per bushel.
**LOAM, FIBROUS.** 3/6 per bushel.
**PEAT, BEST ORCHID.** 7/6 per bushel.
**PEAT, ORDINARY POTTING.** 5/- per bushel.
**SILVER SAND.** Fine and coarse, 7/6 per bushel.
**SPHAGNUM MOSS.** 6/6 per bushel.

## TOOLS AND IMPLEMENTS.

**TREE PRUNERS.** Complete with four strong bamboo handles, each 5 ft. long, can be joined together making two 10 ft. lengths. Complete with pruning saw and pruner for pruning tall trees, price 30/-.
**STANDARD TREE PRUNER.** 6 ft., 9/6 ; 8 ft., 10/6 ; 10 ft. 11/6.
**EDGING IRONS, CAST STEEL.** Handled, 6/-ea.
**FORKS, HAND.** 2/6 and 4/6 each.
**FORKS.** Best quality digging, 4-prong, 6/6 ; 5-prong, 10/- each.
**FORKS.** Ladies' size or Border Forks, 7/6 each.
**REELS, GARDEN.** 4/6 each ; Lines (30 and 60 yds. long), 2/9 and 5/8 each.
**SAWS, PRUNING.** 4/6 and 6/6 each.
**SHEARS, GRASS OR HEDGING.** 9/6 per pair.
    ,,  LOPPING. Very powerful. 27/- pair.
    ,,  PRUNING. 6/6 per pair.
**SHOVELS.** Improved London, 7/6 each.
**SPADES.** Best quality "Norfolk" pattern, 10/6 each.
**SPADES.** Ladies' size, 7/6 each.
**TURFING IRONS.** Best solid blades, handled, 22/-.

## STAKES & TYING MATERIALS.

**BAMBOO CANES.** 4 ft. Thin, 6/- ; Medium 6/6 ; Thick, 12/- per 100.
    ,,  ,,  5 ft. 17/6 per 100.
    ,,  ,,  6 ft. 20/- per 100.
**DAHLIA OR ROSE STAKES.** Painted Green, with tarred ends. 4 ft., 6/6 ; 5 ft., 7/6 ; 6 ft., 9/- per doz.
**LABELS, WOOD.** Painted, in boxes of 100. 4 in., 1/6 ; 5 in., 1/9 ; 7 in., 2/3 ; 8 in., 2/6 ; 9 in., 2/9 ; 12 in., 3/-.
**RAFFIA.** Best quality. Per lb., 1/6
**TAR TWINE.** In balls, thin, medium, and thick, ¼ lb., 1/- ; 1 lb., 2/-.
**TARRED YARN IN HANKS.** 2/- per lb.

## GARDENING GLOVES.

**STRONG TAN.** Suitable for hedging and general gardening. 4/6 per pair.
**MEN'S GARDENING.** Well made, handsewn. 4/6 per pair.
**LADIES' GARDENING GLOVES.** Handsewn. 4/- per pair.
**WOMEN'S TAN GLOVES.** 3/- per pair.
**RUBBER GLOVES.** For use when using insecticides. 12/6 per pair.
Postage on Gloves, 3d. per pair.

## THERMOMETERS.

**THERMOMETERS, BOXWOOD.** 2/9 & 3/9 ea.
    ,,  WHITE JAPANNED SCALES ; especially adapted for the garden (Nogretti and Zambra), 7/6 each.
**THERMOMETERS, MINIMUM AND MAXIMUM REGISTERING.** 12/6 and 16/6 each.
**PLUNGING THERMOMETERS,** for hot-beds, 10/6 each.

## GENERAL SUNDRIES.

**BUDDING KNIVES.** 204, 6/3 ; 204B, 9/- ; 207, 8/- ; 316, 7/6 ; 325½, 9/9 ; 329, 5/9.
**PRUNING KNIVES.** Various sizes. 6/-, 7/6, and 6/3 each.
**SECATEURS.** French pattern, most useful. 6½ in. 6/9 ; 7½ in., 9/9 per pair.
**BROOMS, BIRCH.** 9d. each, 9/- per doz.
    ,,  BASS. 4/6 each.
**LABELS, ZINC, IMPERISHABLE.** For Roses and Fruit Trees. 3,6 and 6/6 per 100.
**METALLIC INK.** For writing on above. 6d. and 1/- per bottle.
**LABELS, ACME GARDEN.** For Roses 2/6 per doz. Fruit Trees 3/6 per doz., postage on 1 doz., 3d. *Please give names of varieties wanted.*
**WEED KILLER, McDOUGALL'S NON-POISONOUS.** 1 gallon tin, 6/6, 5 gallon drums, 25 -.